HEREDITY AND POLITICS

HEREDITY AND POLITICS

J. B. S. HALDANE, F.R.S.

**WELDON PROFESSOR OF BIOMETRY
IN THE UNIVERSITY OF LONDON**

New York

W·W·NORTON & COMPANY·*Publishers*

PRINTED IN THE UNITED STATES OF AMERICA
FOR THE PUBLISHERS BY THE VAIL-BALLOU PRESS

Contents

Preface

IN THE last fifty years we have learned something about human biology, and particularly about human inheritance. This knowledge has yet found little application in Britain. But it has been used to support proposals for very drastic changes in the structure of society. And the stringent measures which have been taken in Germany, both for the expulsion of Jews from many walks of life, and for the compulsory sterilization of many Germans, are said to be based on biological facts.

In the United States there has been a good deal of legislation conferring very wide powers of sterilization on officials. Outside California this has led to little action, thanks, perhaps, to the common-sense of the American people, and the fact that America is much more democratic than England, let alone Germany. American readers may resent any criticism of their institutions by a European, in view of the mess into which Europeans have got themselves.

I have deliberately chosen examples of what I consider to be abuses of the sterilization laws from the United States rather than from Germany, not from any anti-American sentiment, but because I wished to persuade British readers that such abuses may occur under a legal system based on English law, and carried out under the criticism of a press somewhat freer than our own. Other examples of the alleged abuse of these laws, for example the Cooper-Hewitt case, will be familiar to American readers.

I do not believe that our present knowledge of human heredity justifies much of the legislation which is supposed to be based on it. I shall doubtless be accused of allowing my political opinions to override my scientific judgement. It is therefore worth pointing out that the questions with which I shall deal cut right across the usual political divisions. For example, the English National Council of Labour Women has recently passed a resolution in favour of the sterilization of defectives, and this operation is legal in Denmark and other countries considerably to the "left" of Britain in their politics.

It may well be that an increase in our knowledge will fully justify the application to man of certain measures which have led to improvements in the quality of our domestic animals. As one who is endeavouring to increase this knowledge, I can even say that I hope that it will do so. But I believe that the facts concerning human heredity are far less simple than many people think them to be. And I hold that a premature application of our rather scanty knowledge will yield little result, and will merely serve to discredit the branch of science in which I am working.

This book is based on the Muirhead Lectures given at Birmingham University in February and March 1937. I was particularly glad to deliver them owing to my late father's long association with the university. The lectures in question are supposed to deal with political philosophy. If I am accused of deviating from this subject I can plead that Plato was perhaps the first to deal with eugenics, and that as a matter of fact a full discussion of the origins of human inequality raises a number of quite subtle logical problems. And since Professor Muirhead, in whose honour these lectures are given, is perhaps best known to the general public as the author of *The Platonic Tradition in Anglo-Saxon Philosophy*, it is not entirely inappropriate that I should deal with one of the very many problems which Plato raised, and about which we are still disputing.

The first half of the book is mainly devoted to an exposition of the principles of genetics, so far as they apply to men and women. I have tried to make it as popular as possible without sacrificing truth to simplicity. And I do not think that many biologists will disagree with the statements there made. In the second part I deal with more controversial topics. Many people will disagree with my conclusions. I hope that

9

those who do so will point out in detail where, in their opinion, my argument is faulty. It is only by such a dialectical method that we are likely to arrive at the truth on this very difficult question.

HEREDITY AND POLITICS

The Biology of Inequality

I PROPOSE in this book to examine certain suggested applications of biology to political science. In particular I wish to examine certain statements regarding human equality and inequality, some of which have been used to justify not only ordinary policy but even wars and revolutions. In the first chapter I shall quote some statements of five doctrines, and most of the remaining chapters will be devoted to their examination.

We will first consider the doctrine of the equality of man. I will quote from a great revolutionary document of the eighteenth century, the American Declaration of Independence, which was published in 1776 and is mainly due to Jefferson.—"We hold these truths to be self-evident, that all men are created equal, that they are endowed by their Creator with certain unalienable Rights, that among these are Life, Liberty, and the pursuit of Happiness." This or a very similar doctrine of equality was important for the French Revolution. What did it mean in practice? The thirteenth and fifteenth amendments to the United States' Constitution were needed to abolish Negro slavery and racial discrimination in the matter of the franchise. For whites it meant a very considerable measure of equality before the law, and it has, I think, meant a somewhat greater equality of opportunity than exists in England; but it did not give rise to any systematic attempt to bring about economic equality.

Modern revolutionary theory is much more modest in its statements regarding equality, though its practice goes somewhat further in that direction. "The real content of the proletarian demand for equality is the demand for the abolition of classes. Any demand for equality which goes beyond that, of necessity passes into absurdity." So wrote Engels, and the passage was considerably amplified by Lenin. Modern revolutionary theory looks forward to two types of society; Socialist society in which each citizen works according to his ability and receives in accordance with the amount of work done, and Communist society in which each works according to his ability and receives according to his needs. There is a certain approximation towards Socialist society in the Soviet Union, but Communist society remains an ideal. Neither of these theories is equalitarian. Stalin in a report to the Seventeenth Congress of the c.p.s.u. said: "Marxism starts out with the assumption that people's tastes and requirements are not, and cannot be, equal in quality or in quantity, either in the period of Socialism or the period of Communism." Further, so far as I know, official Communist theory includes no clear statement of the origins of inequality other than economic.

Now although Jefferson regarded the truth of human equality to be self-evident there is remarkably little positive evidence for the Jeffersonian theory, and its interest is, I think, mainly historical. We shall have to consider later how much of it can be salvaged. Conservative and reactionary politicians and biologists today lay considerable stress on human inequality.

Let us now consider a series of doctrines that are based on the theory of inequality. We will take first the theory that "the unfit should be sterilized." I may add at once that the operation of sterilization is not castration. It means an operative interference which prevents the conception or begetting of children. It is a slight operation in men, more serious in women. There have been many statements of this doctrine. For example, Mr. Justice Holmes of the United States Supreme Court, in a judgement on appeal said: "It is better for all the world if society can prevent those who are manifestly unfit from continuing their kind." We must ask, however, "Who are the unfit?" and "Do they all continue their kind?" We must also ask who is to decide these questions, both the question of the unfitness and that of whether it is handed on; and we must ask a final question—whether sterilization is the only practicable way of preventing the individual from continuing his kind, if we find that this is desirable. There have been, of course, many attempts to answer this question, and to put sterilization on a legal basis.

I prefer not to quote the German law on the subject because it is inevitable that to do so would give rise to a certain amount of prejudice either for or against this law. I will quote the American model Sterilization Law, drafted by H. H. Laughlin in a *Report of a Psychopathic Laboratory of the Municipal Court of Chicago* (1922).[1] Here are some sections of this law; [Section 2, subsection (a)] "A socially inadequate person is one who by his or her own effort, regardless of etiology or prognosis, fails chronically in comparison with normal persons, to maintain himself or herself as a useful member of the organized social life of the state; provided that the term socially inadequate shall not be applied to any person whose individual or social ineffectiveness is due to the normally expected exigencies of youth, old age, curable injuries, or temporary physical or mental illness, in case such ineffectiveness is adequately taken care of by the particular family in which it occurs."

"(b) The socially inadequate classes, regardless of

[1] Pp. 446, 447 of *Eugenical Sterilization in the United States.*

16

etiology or prognosis, are the following: (1) Feeble-
minded; (2) Insane (including the psychopathic);
(3) Criminalistic (including the delinquent and way-
ward); (4) Epileptic; (5) Inebriate (including drug
habitués); (6) Diseased (including the tuberculous,
the syphilitic, the leprous, and others with chronic, in-
fectious and legally segregable diseases); (7) Blind [1]
(including those with seriously impaired vision); (8)
Deaf [2] (including those with seriously impaired hearing); (9) Deformed (including the crippled); and
(10) Dependent (including orphans, ne'er-do-wells,
the homeless,[3] tramps,[3] and paupers [3])."

"(f) A potential parent of socially inadequate offspring is a person who, regardless of his or her own
physical, physiological or psychological personality,
and of the nature of the germ-plasm of such person's
co-parent, is a potential parent at least one-fourth of
whose possible offspring, because of the certain inheritance from the said parent of one or more inferior
or degenerate physical, physiological or psychological
qualities would, on the average, according to the demonstrated laws of heredity, most probably function as
socially inadequate persons; or at least one-half of
whose possible offspring would receive from the said

[1] E.g. Milton.
[2] E.g. Beethoven.
[3] E.g. Jesus.

17

parent, and would carry in the germ-plasm but would not necessarily show in the personality, the genes or genes-complex for one or more inferior or degenerate physical, physiological or psychological qualities, the appearance of which quality or qualities in the personality would cause the possessor thereof to function as a socially inadequate person under the normal environment of the state."

Now you see that goes rather far! Section 15 of the same draft law empowers the State Eugenicist to cause the potential parents of socially inadequate offspring to be sterilized in a "skilful, safe and humane manner, and with due regard to the possible therapeutical benefits of such treatment of operation." Such a course may be desirable. That is a matter which we shall have to discuss. I do put it forward, however, that such legislation is considerably more revolutionary than, for example, the more moderate forms of Socialism, and would involve considerably more interference with individual liberty. It may be necessary in the interests of the race. That is a matter we shall have to examine later on.

The third statement which we shall have to consider is that certain classes are congenitally superior to others, and that it is desirable that the superior classes should reproduce more rapidly. As an example of that type of thinking I will quote the *Report of the Com-*

mittee of the Eugenics Society (1910) [1] which commented on the *Reports of the Royal Commission on the Poor Laws* published in 1909.—"That element in pauperism which represents and transmits original defect, almost completely neglected in the investigation and wholly neglected in the recommendations of the Commission, is the one we wish to take into consideration. The determination of this element is not a matter of opinion but of the application of methods of careful investigation. It is impossible to disregard the fact that the typical dependent in the minds of the Commissioners is not the typical dependent who habitually receives relief. Yet it is precisely the latter who is primarily the subject of Poor Law relief, and who affords the chief burden on the public purse. He is not the man who responds to a call on manly independence or stands ready to take a place made available through the Labour Exchange. He was born without manly independence [2] and is unable to do a normal day's work however frequently it is offered to him.

"In a general sense, and of course with many exceptions, the unemployed represent relatively weaker stocks. With a diminution of work elimination falls on the less qualified. This is qualified by the factor of age;

[1] Quoted by E. J. Lidbetter, *Heredity and the Social Problem Group*, Vol. 1, p. 12 (1933).

[2] In my own experience the majority of new-born infants are devoid of this quality.

elimination at forty years of age is possibly associated
with elements of original weakness. If a man can do
only half the work required in these days of standardized wages, it is rather futile to attempt to introduce
him to the industrial system."

.   .   .   .

"It is to be noted finally that degenerate tendencies
do not manifest in transmission a single set of characteristics but take on a great multiplicity of forms. A
single family stock produces paupers, feeble-minded,
alcoholics and a certain type of criminals. If an investigation could be carried out on a sufficiently large scale
we believe that the greater proportion of undesirables
would be found connected by a network of relation-
ship; a few thousand family stocks probably provide
this burden which the community has to bear."

It is only fair to remember that that was written in
1910, and that since then the problem of unemployment has entirely changed in character. Those people
who were regarded as unemployable have been called
the "social problem group" by the investigators of the
Eugenics Society. Modern sociologists, however, rarely
ascribe the unemployment of 1½ million people to
congenital abnormalities. We shall later discuss the theory that many of the poor are poor because of hereditary defects.

In the same way many biologists believe in the innate superiority of certain classes and in the extreme importance of the ruling classes. Professor Fisher, for example, in his book, *The Genetical Theory of Natural Selection*,[1] writes: "The fact of the decline of past civilizations is the most patent in history, and since brilliant periods have frequently been inaugurated, in the great centres of civilization, by the invasion of alien rulers, it is recognized that the immediate cause of decay must be the degeneration or depletion [2] of the ruling classes. Many speculative theories have been put forward in explanation of the remarkable impermanence of such classes." We note that Fisher takes for granted the exact opposite of the proposition concerning equality that was obvious to Jefferson. The truth may lie between the two ideas.

We have next to consider the fourth doctrine "that certain races are congenitally superior to others." The earliest statement of that doctrine known to me is found in the Book of Genesis, where the curse on the children of Ham is related. It is worthy of note that if this attribution of priority is accurate, the doctrine of racial superiority is originally a Jewish doctrine, al-

---

[1] P. 237.
[2] Thus if English culture decayed during the sixteenth century and particularly under Elizabeth, this may be attributed to the depletion of the feudal nobility in the Wars of the Roses. If not, not.

though it is now being used against the Jews in Central Europe. There have been some very surprising statements of this doctrine by recent German authors since 1933. I will only quote one to give an example of the remarkable theories current in Germany. Dr. Johann von Leers writes: "After a period of decadence and race obliteration, we are now coming to a period of purification and development which will decide a new epoch in the history of the world. If we look back on the thousands of years behind us we find that we have arrived again near the great and eternal order experienced by our forefathers. World history does not go forward in a straight line but moves in curves. From the summit of the original Nordic culture of the Stone Age, we have passed through the deep valley of centuries of decadence, only to rise once more to a new height. This height will not be less than the one once abandoned, but greater, and that not only in the external goods of life."

It is interesting to think that the Nordic race, if properly purified, may rise even higher than the culture of the Stone Age. When one reads such statements as that, one is tempted to ask whether they are made in order to obtain or retain posts, or whether, possibly, they may not be a rather subtle form of propaganda intended to make the existing racial doctrines in Germany appear ridiculous. I therefore propose to quote from German authors who wrote before the advent to power of the National Socialist Party, and who therefore expressed themselves much more moderately. For example, E. Fischer, writing in 1923, gives the following account of the Nordic race: "The mental endowments of the Nordic race are great energy and industry, vigorous imagination, and high intelligence. Conjoint with these are foresight, organizing ability and artistic capacity (this being least marked in respect of music); and also the unfavourable qualities of strong individualism, a lack of community sense and of willingness to obey orders, a certain one-sidedness and an undue inclination towards imagination and flights of fancy, a dislike for steady and quiet work; while as additional qualities may be mentioned a considerable expansive force, a power of devotion to a plan or idea, an adequate capacity for instilling an idea into others and a small inclination for adopting the ideas of others —in a word significant powers of suggestion and comparatively little suggestibility. It is obvious that when circumstances are favourable persons richly endowed with these gifts are likely to become leaders, inventors, artists, judges and organizers." These admirable qualities are regarded as inherent in the Nordic race, and it is thought that the regeneration of Germany can only come from the spread of the Nordic elements in it. For example, Günther in 1929 wrote: "The rise of the

23

peoples of Germanic speech is given by the increase of the healthy hereditary units, and an increase of Nordic blood."

Later, however, there has been a tendency to speak not so much of the Nordic race but of the German race. We shall have to ask whether a Nordic or German race exists, and if so whether these doctrines as to natural endowments can be justified.

The fifth doctrine is that "crossing between different races is harmful." Lenz wrote: "In my opinion there can be no doubt that the mingling of races widely distinct from each other may lead to the production of types which are disharmonious in respect alike of body and mind." More recent German statements of the terrible effects of race mixture will be familiar to you. As the most striking of them, those of Herr Streicher in *Der Stürmer*, are highly obscene, I do not propose to prejudice the minds of readers by quoting them.

Now before we can examine these and related theories in detail we must consider the general biology of inequality. That is dealt with by the science of genetics. Genetics is primarily concerned with innate inequality, but has to consider all kinds of inequality, or as we call it in biological terminology "variation." The success of genetics, which deals among other things with heredity, has been due to the fact that during the last forty years it has been concerned with differences rather than with resemblances. It is possible to give a reasonable answer to the question, "Why is this mouse black while this other mouse is white?" but at present it is not possible to give at all an adequate answer to the question, "Why are they both mice? Why does a pair of mice produce another mouse and not a rat or a motor-bicycle?" By concentrating on these relatively small differences genetics has advanced a considerable distance, and it is therefore peculiarly adapted to deal with this problem of human inequality. I am certainly not going to answer such a question as "What is Man?" I may be able to help you to get a clearer view as to the nature of the differences between individual men and their causation. If so I may be able to answer at least provisionally some of the questions which I have posed already.

Let us suppose that we have before us two dogs, both with legs somewhat bent. It may be that one of these dogs has bent legs because as a puppy it received a diet containing inadequate quantities of anti-rachitic vitamin, while the other dog has bent legs because its father was a dachshund. In the first case we say that the difference between this dog and a straight-legged dog is due to nurture. In the second case we say that it is due to nature. It is possible by experimental methods with animals and plants to separate these two causes

25

of variation, but we must recognize that in the majority of cases both of the causes are operative. If we are dealing with the sizes or weights of a number of animals we shall certainly find that the differences are due both to differences of nature and to differences of nurture. In man, where experiment is impossible, we shall find it very much harder to determine the origin of the differences that undoubtedly exist. Let us see what happens when we tackle the problem experimentally.

If in any branch of science we find that one quantity or quality varies as a function of several others we shall design our experiment so as to keep all but one of the independent variables approximately constant. If, for example, we wish to ascertain the laws governing the volume of a gas we shall first keep the temperature constant while varying the pressure and so discover Boyle's law. If we keep the pressure constant, while varying the temperature, we shall discover Charles's law. If we begin by measuring the volume of gas at an arbitrary series of temperatures and pressures we shall find our work very much more difficult.

Now we do in practice try to eliminate our variables —our differences of nature and of nurture. I have not defined them because we shall be able to understand their essence very much more clearly when we deal with the practical methods that render them uniform. The most obvious thing to do is to make the nurture of our various organisms as like as possible. If we are growing a number of plants we shall see that they all have the same soil, the same amount of water and light, and that the density of plants in one part of our field is the same as in the other. With animals we shall take similar precautions. For example, we shall be particularly careful that if there is any infection all members of the population shall be equally exposed, and if we were dealing with man in an ideal experiment we should have to try to render the education and social environment of our different individuals as uniform as possible. We should then see that any differences that remained in that uniform environment were probably due to nature and not nurture.

I am perfectly aware that a uniform environment is an impossible ideal, nevertheless it is easy to think of characters which are very slightly affected by the environment, for instance, eye-colour in man. It is easy to think of other characters, which, although they vary considerably with the environment, may quite readily be stabilized by making the environment similar—for example, the skin colour which varies a good deal according to the amount of sunlight to which a man is exposed. We may take it that during a winter in England there will not be very much sun-burn.

We see then that it is possible to a large extent to eliminate one of our variables, nurture, at least when doing experimental work with plants and animals. How shall we perform the converse operation? How shall we get a population of animals or plants uniform as regards their nature, their innate qualities? There are three ways of doing so. First of all we may grow what is called a clone, that is to say a population of individuals which are derived from the same individual by vegetative reproduction. For example, if you buy a named variety of a potato, tulip, rose, or apple, you will find, if your seedsman is an honest man, that the plants which you buy have all been derived from the same seedling by vegetative reproduction. The original potato was divided; it sent out roots on which new tubers were grown, and these were used as so-called seed potatoes for the next generation, but there was no sexual reproduction. If we take a well-known type such as Arran Victory we shall find that it produces a large variety of different potatoes if its seeds are sown. In the case of a named rose we propagate it by grafting but there is no sexual reproduction. Within a clone we find considerable uniformity, and in so far as there are differences they are not in general handed down. Selection within a clone is ineffective. If you once have your named variety of rose, except for a very occasional bud sport, you will not improve it by selecting the best individuals from it. Such differences as exist appear to be temporary effects of environment which are not transmitted.

One may ask, "What has that to do with man? Man does not reproduce vegetatively." In human beings there are two types of reproduction, the ordinary sexual reproduction, and much more rarely asexual reproduction. The embryo in its early stages may divide to give a pair of monozygotic twins, who resemble one another to a very remarkable degree, and who are believed, on genetical grounds that seem to me entirely sound, to have the same nature. They have, of course, very much the same nurture up to the time of birth and often afterwards.

A second type of genetically uniform population is that called a "pure line." You may get that by self-fertilizing a plant for ten generations, or in the case of animals by brother and sister mating, for a larger number—thirty or more—of generations. Such a pure line is generally very uniform. There may, however, be differences within it.

The remarkable point is that these differences are not inherited. It is easy to breed pure lines of the fly called *Drosophila funebris*, since it accomplishes a generation in about twenty days, and one can breed 400 in a single bottle. In the normal type of fly the veins extend to the margin of the wing. In some abnormal individuals one is broken. It is possible by suitable crossings to produce populations in which a given proportion of the individuals have that vein interrupted. It may be very few; it may be 100 per cent; it may be intermediate. It is possible by continued brother-sister mating to obtain a pure line in which the proportion of abnormals is the same in all families.

Table 1 gives the actual figures obtained by Timoféeff-Ressowski in such a line.

TABLE 1

Offspring of Abnormal Flies		Offspring of Normal Flies	
Abnormal	Normal	Abnormal	Normal
199	24	311	36
288	32	201	25
192	22	219	25
679	78 $= 10\cdot 3 \pm 1\cdot 1\,\%$	731	86 $= 10\cdot 5 \pm 1\cdot 0\,\%$

It will be seen that in this line 10 per cent of the individuals had normal wings, and 90 per cent abnormal. These proportions were the same whether he bred from two parents with normal wings or two with broken veins, provided both were members of the line. Selection for normality or abnormality is completely ineffective. That is to say, although there are differences with regard to the wing vein those differences are not inherited. What is inherited is a constitution such that in a particular environment 10 per cent have a normal wing and the remainder have a broken vein. I am aware that that is a somewhat difficult conception to grasp. It is fundamental in modern genetics. We cannot always speak of the inheritance of a character; in many cases we can speak of the inheritance of a constitution which in a particular environment will give such and such a range of characters.

Now within a pure line all differences, as far as we can see, are due to nurture, none to nature. If we alter the conditions so that a larger proportion of our flies have a particular character, that character is not handed on to the offspring when the original environment is restored. As a result of such experiments very few geneticists nowadays believe in Lamarck's doctrine that "acquired characters are inherited."

The third type of genetically uniform population which we can get is the first cross between two pure lines. A second generation is generally very variable, but the first cross is often uniform and considerably more vigorous than either of the lines.

A study of pure lines teaches us that there is a certain residual variation which we cannot eliminate, even if we eliminate all differences of heredity. It is possible that if we could get an absolutely uniform environment we could eliminate these differences also. In an environment as uniform as we can get we shall still find them.

In an ordinary population, for example in any human population, there are no pure lines—a point of very considerable importance. A pure line, however, is not merely a laboratory curiosity. The named varieties of many seed-plants, for example, wheat or sweet peas, approximate very closely to pure lines. Although, therefore, the pure line has no immediate applicability to human problems it can give us a great deal of information. For example, we are apt to think of congenital qualities in a baby, qualities with which it is born, as being probably due to nature, likely to depend on the make-up of its parents, and likely to be transmitted to the offspring. Let us see how far that is true by considering a particular character as manifested in four pure lines of guinea-pig.

Guinea-pigs quite frequently have extra toes. By suitable selection combined with inbreeding it is possible to produce a pure line in which the frequency of extra toes may vary from 12 per cent to 56 per cent in particular cases. You will not get any more extra toes from the extra-toed members of one of these lines than from the normal ones. The percentage of extra toes represents the reaction of that line to its particular environment. We next ask what is the most important element in the environment determining extra toes. In the particular set of environments met with in Wright's work by far the most important element was

TABLE 2

Percentages of Polydactylous Guinea-pigs

Age of Mother	Line A	Line B	Line C	Line D
3–6 months ..	29·3	34·6	68·1	81·0
6–9 months ..	7·4	28·2	54·4	69·5
9–15 months ..	9·6	21·9	28·9	50·0
15– months ..	6·1	12·1	22·0	30·2

Effects of heredity and environment on the frequency of polydactyly (extra toes) in 1,986 guinea-pigs. After Wright.

the age of the mother. You will see that in line A the young mothers produced 20 per cent of offspring with extra toes and the old mothers only 6 per cent. In line D the young mothers under six months produced 81 per cent and the older ones over fifteen months only 30 per cent. This at once shows that a character can be determined to a considerable extent both by heredity and by environment. The differences between these four lines are, of course, hereditary. The differences between the different rows in the diagram are environ-

33

mental. If one is a rabid environmentalist one will read that table from top to bottom, if a rabid eugenist, from side to side. If one is a biologist one will read it both ways.

One type of human mental defect is determined in this way, namely, "Mongolian imbecility," a condition in which, as Penrose has conclusively shown, the age of the mother is an important determining factor. The average age of the mothers of these imbecile children is about thirty-nine years, whilst that of mothers of normal children is under thirty. Besides this environmental factor there is a genetical factor, as is shown by the fact that two mothers of such imbeciles are often related. This means that an embryo of a certain constitution will develop into a Mongolian imbecile in a particular type of prenatal environment provided by an elderly mother, or that a mother of a certain constitution provides a special type of prenatal environment when she ages. There is some evidence that a few other kinds of mental defect are determined in a similar manner. On the other hand mental defect due to injury at birth seems to be commoner in first-born children, who are generally brought into the world with more difficulty than later children.

It is frequently asked, "What is the relative importance of nature and nurture?" That is a question to which no general answer can be given. It is obvious that if in the population of guinea-pigs no female were allowed to breed until she was six months old the differences due to nurture would be considerably reduced. If the population had contained only three lines instead of four the differences due to nurture would have been diminished. It is possible, by suitable choosing of your character, your population, and your environment, to produce a population in which a given character is determined entirely by differences of nature or entirely by differences of nurture, and therefore the question has to be answered separately for any given population and any given character. For example, we may take such a character as illiteracy and we may compare the amount of illiteracy in adults under forty in England and India. In England we should find that the people who could not read were almost all either blind or mental defectives. We should find reason to believe that a considerable amount of the blindness and mental defect in England was due to differences of nature. On the other hand, if we went to India we should find that the majority of the illiterates were illiterate because they had had no opportunities for learning to read, and therefore differences in that respect were mainly matters of nurture. One could give, of course, many more examples of the same kind. The important point is to realize that the question of the relative importance of nature and nurture has no general answer, but that it has a very large number of particular answers.

It is fortunate for our purpose that although pure lines do not exist in man there are nevertheless human groups which breed true, or very nearly true, for certain physical characteristics. For example, we can be reasonably sure that the skin colour of the children of two English people will vary between fairly narrow limits, while the children of Negroes will vary between other limits, but there will be no overlapping. We shall have to consider whether there is evidence for the existence of psychological characters that are equally closely determined.

Before we do that we must consider the interaction of nature and nurture. Let us suppose we have two different stocks which are pure lines or at least do not have very great innate variation as regards the particular character which we are studying. It may be a physical character such as weight, a physiological character such as milk yield, or some of the numerous forms of human achievement. But we will suppose that we can order our populations as regards their achievement. We can probably say that this group is on the average significantly heavier than that group. By "significantly" I mean that the difference is such that it cannot well be due to sampling error. We may say with regard to a particular intelligence test, "This group does significantly better than that one," and that is a statement we can make quite regardless of the philosophical question whether intelligence can be measured.

TABLE 3

		X	Y			X	Y
1.	A	1	2	*or*	A	1	3
	B	3	4		B	2	4

		X	Y
2.	A	1	4
	B	2	3

		X	Y
3.	A	1	2
	B	4	3

		X	Y			X	Y
4.	A	1	3	*or*	A	1	4
	B	4	2		B	3	2

Now suppose that we have two races A and B in two environments X and Y. And suppose that we have samples of each race in each environment sufficiently large to enable us to order them without doubt as to their achievement in some respect, say longevity, milk yield, or intelligence. If their achievements overlap, we can still order them with certainty by taking large enough samples.

There are exactly four possibilities, shown in Table 3. The enumeration is so simple that no one has ever troubled to make it. Nevertheless, I believe it is worth making. In the first type of interaction race A is superior to race B in each environment, and environment X is more favourable than environment Y to each race. The numbers, 1, 2, 3, 4 denote the order of achievement of the four populations. This is a common type of interaction. It would be exemplified if we took two races of dog, say mastiffs and dachshunds, as races A and B, and a good diet and a starvation diet as environments X and Y. It is clear that on the better diet each race of dog would be heavier than on the poor diet. But on each diet the mastiffs would be heavier than the dachshunds. If nature and nurture always interacted in this way we could say with scientific accuracy, "This is a heavier race of dog." "This is a more musical race of men." "This is a more fertile breed of poultry." "This is a healthier environment than that." But unfortunately, things are not always so simple in reality.

Now consider the second kind of interaction. Let A be Jersey cattle and B Highland cattle. Let X be a Wiltshire dairy meadow, and Y a Scottish moor. On the English pasture the Jersey cow will give a great deal more milk than the Highland cow. But on the Scottish moor the order will probably be reversed. The Highland

cow will give less milk than in England. But the Jersey
cow will probably give less still. In fact, it is very likely
that she will give none at all. She will lie down and die.
You cannot say that the Jersey is a better milk-yielder.
You can only say that she is a better milk-yielder in a
favourable environment, and that the response of her
milk yield to changes of environment is larger than
that of the Highland cow. Our specialized domestic
animals and plants generally behave in this way. It
is likely that certain human types react in a similar
manner.

It is, of course, possible that the interaction between
nature and nurture is of a simpler type in the determination of human intelligence than in the milk yield
of cattle or the seed yield of wheat plants. But even a
thorough-going materialist might well doubt this. Unless it is true we cannot in general say that A has a
greater innate ability than B. A might do better in environment X, and B in environment Y. Had I been
born in a Glasgow slum I should very probably have
become a chronic drunkard, and if so I might by now
be a good deal less intelligent than many men of a
stabler temperament but less possibilities of intellectual
achievement in a favourable environment. If this is so
it is clearly misleading to speak of the inheritance of
intellectual ability. This does not mean that we must
give up the analysis of its determination in despair. It

means that the task will be harder than many people believe.

The third type of interaction may be illustrated by normal (A) and mentally defective (B) human children. The normal children will do best in an ordinary school (X), but even in a special school (Y) for mental defectives they will do better than defective children. On the other hand, the defective children will do better in the special school. In each environment the normal children will be superior. But the environment which is better for the normal child will be worse for the defective child, and conversely. We could equally well illustrate our case if A were normal bean plants and B a race of beans which turn white in strong light, while X was full sunlight, and Y partial darkness.

As an example of the fourth type let A be Englishmen and B West African Negroes. Let X be an English town and Y the Gold Coast colony. Let the four populations be placed in order of their average lengths of life. We should probably find that the order was: English in England, Negroes in Africa, Negroes in England, English in Africa. We should certainly find that each race lives longer in its native environment than when transplanted. We could not say that as regards health as measured by longevity either race or either environment was superior to the other. The Englishman in West Africa is killed off by yellow fever, the

THE BIOLOGY OF INEQUALITY

Negro in England by tuberculosis, each having a considerable immunity to the disease prevalent in his native land.

If we merely have two races and two environments Table 3 exhausts the possibilities unless two or more of the four achievements are equal. With a number of races and environments things are of course more complicated. But after studying Table 3 we shall be a little suspicious of such phrases as "a good heredity," "a good environment," or "a superior race." Unfortunately, almost all current theory is based on the view that the first type of interaction is universal, and this applies equally to the supporters and to the critics of eugenics.

In a mixed population things are not so simple. We may find populations in which most of the differences are due to heredity in the strict sense of a resemblance between parent and offspring. If you ask why a given dog is a greyhound, it will be correct to answer "because both his parents and his ancestors for some way back were greyhounds." If you ask why a given cat is tabby it will not usually be accurate to say "because both his parents were tabbies." In that respect the cat population presents a closer analogy to the human population than do the dogs. Let us try and see what we have to deal with besides heredity as a cause of innate differences, differences of nature. Suppose we cross a pure bred black rabbit with a pure bred blue rabbit, the hybrids will be black. But if we cross the hybrids together we shall get some blacks and some blues. The differences between the blacks and the blues are differences of nature, because, unlike differences due to nurture, they are handed down to the offspring. The process by which the black rabbits give rise to blacks and blues is called segregation. We shall examine it in greater detail later on. It must be carefully distinguished from the effects of environmental differences which are not transmitted to the offspring. The kinds of differences which we may get within a human population are summarized in Table 4:

TABLE 4

NURTURE.   Differences due to different environments.

NATURE. { Heredity. Differences of ancestry.
Segregation. Differences due to chance combinations of genes.
Mutation. Changes in genes.

X.

Causes of human inequality.

First of all there are differences that are due to differences of nurture—the difference between a sun-burned child and a child that is not sun-burned; and between a normal child and a rickety child. Secondly, there are differences of nature, which fall into three categories; differences of ancestry, for example, the difference between a Negro child and a white child, which is due to heredity; differences between brown-eyed and blue-eyed sibs, and in general all heritable differences between brothers and sisters which are due to segregation; differences due to mutation, a rare event of considerable biological importance. I leave a blank space X for differences which cannot be ascribed to any of these. If there is such a thing as freedom of the will in the more extreme sense that comes under X. I regard it as unscientific to leave out X, if only for this reason, that if there is no such thing as X, if all differences between human beings are strictly determined, then it should be possible in the course of some centuries to prove that, let us say, 99.9 per cent at least of all differences of certain kinds are determined by differences of nature or nurture. To my mind a proof that 99.9 per cent were so determined would be very much more effective than an assertion on *a priori* grounds that 100 per cent were so determined. If therefore we leave X in our table we can say that in certain cases, for example that of skin colour, X is fairly small, and we may hope according to our philosophical views to prove either that X is negligible or considerable as regards differences of conduct.

43

In subsequent chapters I will deal in detail with the various questions which I have raised. The latter part of this first chapter may be regarded as a prolegomenon to any systematic treatment of human inequality.

CHAPTER II

The Principles of Human Heredity, as Illustrated by Certain Hereditary Abnormalities

---

IN THE first chapter we dealt with the determination of the various diversities between individuals by differences in their nature and nurture. The interaction between nature and nurture is exceedingly hard to disentangle, even when experimental methods are available to us; it is almost impossible to disentangle when experimental methods are impracticable.

When we are dealing with congenital conditions we wish to study the laws of heredity, and we shall study those most readily upon characters where the environment has relatively little influence, that is to say, upon characters which are such that a given genotype is similar in a great variety of environments.

Let me explain what I mean by the word genotype; it is a word which we shall be using from time to time. We say that a number of animals possess the same genotype if their breeding behaviour is similar, even though they may not be alike themselves. For example, in the first chapter we dealt with flies belonging to the same pure line, some of which had normal wings and others abnormal wings, but which were all alike in their behaviour as parents. A given genotype therefore, which is believed to represent a certain combination of genes (a word which we shall define later on), is not necessarily alike in all environments, but in our study of heredity we shall try to concentrate as far as possible on genotypes rather than on external appearances. Our task will be greatly simplified if we deal as far as possible with characters on which nature has but little effect.

Now where heredity is clearly understood it is generally, if not always, found to follow the laws which were first discovered by Mendel in 1865. Nonbiologists often find a certain difficulty in following these laws, so I am going to use analogy rather freely.

In Spain every person has two surnames, one of which is derived from the father and one from the mother. For example, if you are called Ortega y Lopez you derive the name Ortega from your father, it was his father's name; and Lopez from your mother, as it was her father's name. Now I want you to imagine a strange savage nation in which everybody has two surnames; and when a child is born there is a curious ceremony by which he receives one of the surnames from one parent and one from the other, these being drawn at random by a priest, whose business it is. For example, if Mr. Smith-Jones marries Miss Brown-

Robinson the children may get the name Smith or Jones from the father and Brown or Robinson from the mother; and will be called either Smith-Brown, Smith-Robinson, Jones-Brown or Jones-Robinson. If the priest draws at random those four types occur with equal frequency.

A complication arises from the fact that some people may get the same name from both parents, and be called Smith-Smith, transmitting the name Smith to all their children. If you try to remember that simple scheme you will find no great difficulty in understanding the laws which govern heredity in a very large number of cases.

The things which are analogous to the surnames are called genes. These are material particles of small size which are found in the nucleus of each cell. The cell nucleus contains bodies which are called chromosomes, and these chromosomes (with an exception to be noted later) occur in pairs, one set being derived from the father and another from the mother.

A given gene occupies a definite place in a definite chromosome. There is, therefore, a possibility of there being two like genes, one derived from the father and one from the mother, in a particular place on the chromosome; in which case the individual is called a homozygote. If the corresponding genes are unlike, he or she is called a heterozygote. The homozygotes are

47

Smith-Smiths; the heterozygotes Smith-Joneses. Homozygotes, as regards a particular locus in the chromosome, will hand down similar genes to all offspring; heterozygotes will hand down two types of gene, each one to approximately half the offspring.

Figure 1 is part of a pedigree of brachydactylism, a condition in which the fingers are short owing to in-

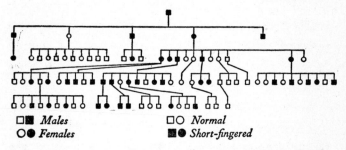

Males

Females

Normal

Short-fingered

FIG. 1.—*Part of a pedigree of short fingers; after Drink-water.*

adequate growth of the middle bone. The long bones of the leg are also short, so that stature is low, though the abnormal people cannot be called dwarfs. The hands are clumsy, but the general health is good. In fact, the brachydactylous people appear to be rather healthier than their normal brothers and sisters.

I have abridged the original pedigree in two respects. In the first place I have omitted a number of normal descendants of normal people. And I have also left out the husbands and wives, all of whom were normal. The

48

two families on the left hand in the fourth generation were by different fathers. Several things are noticeable about this pedigree. In the first place an affected person transmits the abnormality to about half his or her children. Actually in this pedigree it was transmitted to 38 out of 69 children. If we tossed a coin 69 times, we should expect as large a deviation from equality in 47 per cent of trials. So the deviation is not significant. It is also transmitted both by men and women, and each transmits it about equally to their sons and daughters. It never skips a generation. That is to say it is never transmitted by a normal person.

A character which behaves in this way is said to be due to a dominant gene. To use our analogy of names, all the normal members of the population are Smith-Smith. The abnormals are Smith-Jones; they hand down "Smith" to half their children, and "Jones" to the less fortunate half. Smith-Jones can always be distinguished from Smith-Smith. Strictly speaking, a dominant character is one in which Jones-Jones and Smith-Jones (the homozygous and heterozygous dominants) look alike. In most human cases we do not know what Jones-Jones would be. In the case of a slight affection of the fingers in a Swedish family the union of two affected cousins gave a child without arms or legs. Possibly a homozygous brachydactyl might be equally abnormal, or even incapable of life. Nevertheless we shall use the word "dominant" in this rather loose sense to denote characters due to genes which show up in the heterozygous condition.

Fig. 2 is a pedigree of lamellar cataract, present from birth. The inheritance is the same as in the last case.

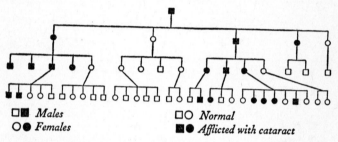

□■ *Males*
O● *Females*
□O *Normal*
■● *Afflicted with cataract*

FIG. 2.—*Pedigree of lamellar cataract; after Harman.*

Again normal spouses are omitted, as are children who died in infancy before their eyes were examined. As here presented the pedigree is perfectly straightforward. But the facts were not so simple. The cataract varied from a large opaque body occupying most of the lens, and causing nearly total blindness, to a small white opacity lying in front of the lens. The father of the five children on the left in the fourth generation had good sight, and was only found to have slight cataract when the family was being investigated. On the other hand, one of his sons was so badly affected that his lenses had to be removed in childhood. Thus a cursory examination of this family would have led to

50

the conclusion that the abnormality sometimes skipped a generation.

Fig. 3 is a pedigree of retinitis pigmentosa, a disease causing partial or complete blindness. This pedigree

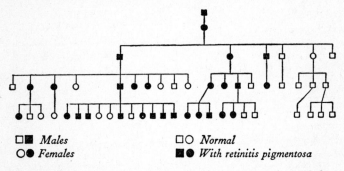

□■ *Males* 　　　　　 □○ *Normal*
○● *Females* 　　　　　 ■● *With retinitis pigmentosa*

FIG. 3.—*Pedigree of retinitis pigmentosa, causing night-blindness in youth, and almost complete blindness at about 40. Normal spouses and some normal children of normals omitted. The gene concerned is partially sex-linked.*

represents a slightly different condition in that the character is not handed down impartially as regards sex. A man who gets it from his father will hand it down predominantly to sons. A man who gets it from his mother will hand it predominantly to daughters. There have been exceptions, however. For example, number 5 in the fourth generation although he received it from his father handed it to two daughters.

I mention this case as showing a rather unusual type of heredity which enables us to locate the gene in question at a definite point on a definite chromosome, whereas in most cases in man we are still unable to say on which particular chromosome a particular gene is carried.

You will notice that among the pedigrees there are two of eye disease, and in almost all collections of pedigrees you will find a large number referring to abnormalities of the eyes. The reason for this is simple and interesting. The eye is one of the few organs which can be really satisfactorily examined, by means of the ophthalmoscope. The doctor is not content to say that a given man is blind or has bad sight. He can examine the eyes very completely. He can state in what part the defect is located. It may be a defect of the cornea, a defect of the lens or the retina, or merely a defect in the shape of the eye. With no other organ is anything of the kind possible. One can draw up pedigrees of heart disease, but in many cases the exact diagnosis is somewhat uncertain. We cannot classify heart diseases yet as we can classify eye diseases because the heart can only be thoroughly examined after death, at a time when it has already ceased to function. In the case of brain diseases we are still almost completely in the dark as regards classification.

Now if we attempted to discuss such a topic as the inheritance of blindness we should be unable to say very much. We should find that in some cases blindness was hereditary and in others it was not; that sometimes it was inherited in a relatively simple manner, and sometimes in a very complicated manner. It is only because we can recognize several hundred clinically different types of eye trouble, of which a certain fraction—and only a certain fraction—is hereditary, that we are able to deal with the matter at all scientifically. This is of some importance, because later on we shall have to discuss the inheritance of various kinds of mental defects. And our conclusions on the subject will be far from clear-cut. I have no reason to doubt that in the course of time we shall be able to deal with that topic as clearly as with the inheritance of eye defects, but that will only become possible when we are able to classify the various types of mental deficiency clinically as we can now classify the various types of defective sight.

The genes whose action I have so far demonstrated belonged to the class which are called dominant. That is to say the heterozygotes differ from the normal homozygotes. In other words, Smith-Jones can be distinguished from Smith-Smith, but in no case do we know what Jones-Jones would be like, that is to say, what would be the homozygote from one of these abnormal genes. It is quite likely that such a person would be extremely abnormal; he or she would possibly have no eyes or legs at all.

A large number of human abnormalities are inherited in this straightforward way. The following is a list of some of the better-known abnormalities;

EYES
> Juvenile cataract (opacity of lens).
> Glaucoma (high pressure in eye).
> Night-blindness.
> Retinitis pigmentosa, etc.

SKIN AND HAIR
> Piebaldness (some types).
> Telangiectasis (bleeding from nose and elsewhere).
> Tylosis (thick soles and palms).
> Dystrophy of nails (about 6,000 cases in French Canada).
> Neurofibromatosis (fairly fatal, mutation frequent).
> Epiloia (tumours in skin, brain, etc.), etc.

BONES AND TEETH
> Lobster claw, or split hand.
> Brachydactyly (short fingers, short stature).
> Cleidocranial dysostosis (abnormal skull and collarbones).
> Absence or smallness of lateral incisors.
> Defective enamel, etc.

54

NERVOUS AND MUSCULAR SYSTEM
  Huntington's chorea.
  Tremor hereditarius.
  Peroneal atrophy (some types).

Many of these may also be due, in some cases, to recessive genes or environmental effects.

We have next to deal with a different class of hereditary abnormalities: those due to recessive genes. Keeping to our analogy in these cases, if Smith-Smith is the normal type, Smith-Jones is indistinguishable from it, and only Jones-Jones is abnormal. In other words, for the abnormality to show, an individual must receive abnormal genes from both parents.

A disease of this type is juvenile amaurotic idiocy. The victims of this disease are born to all appearance normal. Round about 5 years of age they start to go blind. Blindness is fairly complete by the age of 8, and meanwhile the mental powers deteriorate. By the age of 15 the patients are completely idiotic, and later they waste away, and are generally dead before their 20th year.

This disease is to be distinguished from another very similar disease in which the onset is in infancy. It is undoubtedly the more distressing of the two for the parents, for the child is apparently quite normal until

55

the age of 4 to 8, and then gradually dies after many years of illness.

Fig. 4 is a pedigree of the disease; it is one of some fifty pedigrees compiled by Sjögren in a monumental study of the relatives of every single child in Sweden who had that disease. He went through all the schools for blind children in Sweden looking for cases, and spent three years studying their ancestry, favoured by the fact that, in Sweden, family records are kept by churches and go back to the eighteenth century.

The pedigree is typical. Four individuals suffered from the disease; the remainder were normal. It is at once clear that that disease is not in the ordinary sense of the word a hereditary one. It could not be so because individuals who have it die before they can produce children. Nevertheless it is hereditarily determined.

It will be noticed that both pairs of parents were themselves blood relations. That is entirely characteristic of recessive abnormalities, and it is easy to see why. Calling these children Jones-Jones, the parents were each Smith-Jones, and the abnormal gene which we call Jones has come down from one or the other of the two ancestors in the first generation by two different routes. In each case the parents were heterozygous individuals who were themselves at any rate sufficiently normal to marry and earn livings, and it was only because these heterozygotes had the misfortune to intermarry that they produced idiotic children.

Such abnormalities due to recessive genes satisfy a variety of criteria. In the first place they are very much more frequent among the offspring of related parents

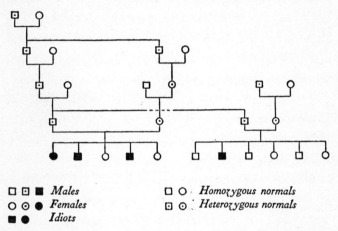

□ ⊡ ■ *Males*   □ ○ *Homozygous normals*
○ ⊙ ● *Females*   ⊡ ⊙ *Heterozygous normals*
■ ● *Idiots*

FIG. 4.—*Pedigree of juvenile amaurotic idiocy; after Sjögren.*

One or other of the spouses in the first generation was heterozygous.

than in the general population; secondly, they tend to occur in sibs; thirdly, they tend to occur in the progeny of related parents, here in the progeny of two brothers; and finally their frequency in families satisfies certain statistical laws, with which I will not deal here.

The interpretation of such pedigrees as this was only

possible as a result of experiments on plants and animals, which made the nature of recessive abnormalities perfectly clear.

Now on the basis of Sjögren's survey of Sweden we are able to come to certain conclusions. The frequency of that particular abnormality was approximately four per 100,000 children born. Among the parents of these children no less than 15 per cent were first cousins and another 10 per cent were blood relations of some kind or another, whereas in the general population only 1 per cent of all marriages, approximately, were between first cousins.

One can make a further calculation. One can ask what proportion of the entire population are heterozygotes? That is to say, what proportion carries this abnormal gene "Jones" which has the property that an individual carrying one is substantially normal, while one carrying two such is destined to blindness and idiocy? The calculation shows that the frequency of these genes in the population is approximately half per cent. That is to say, approximately 1 per cent of all individuals in Sweden are heterozygous for that gene. That is a surprisingly high proportion. In the absence of inbreeding one marriage in 10,000 would be between persons carrying that gene and one in four of their children on the average would be idiots. One would expect therefore to get one idiot among 40,000 children if the marriage of heterozygotes were strictly at random. You get somewhat more than that (about 1 in 30,000) because marriages between blood relations are very much commoner than they would be if a given Swede was equally likely to marry any member of the other sex in the whole of Sweden. And that inbreeding, though slight in degree, accounts for about 25 per cent of all the idiots of this particular type in Sweden.

There are a good many conditions which behave in a similar manner, as recessives. For example, the infantile type of amaurotic idiocy which generally kills children round the age of two behaves in this way. So does albinism. So does retinitis pigmentosa in some pedigrees.

So also does a peculiar and interesting condition discovered by Föllings and called phenylketonuria. He noted that about 2 per cent of the idiots and imbeciles in a certain institution secreted in their urine phenylpyruvic acid, which is a product of incomplete oxidation of the well-known protein constituent phenylalanine. Penrose in England studied the matter further and concluded that the condition is recessive, although not completely so.

No mentally normal individual has ever been found to secrete that substance in the urine, and it is fairly clear that this particular metabolic abnormality is responsible for a certain percentage, not more than 1 or 2 per cent, of all mental defect.

Where it is segregated it is exceedingly clear-cut. One family investigated by Penrose contained two hopeless idiots, chemically abnormal; the remaining five sibs were all mentally as well as biochemically normal, one so able as to win the second best scholarship in his year in a fairly large English county. The demarcation between the normals and abnormals was absolutely sharp.

One particular abnormality which is, very largely at any rate, due to recessive genes, is congenital deaf mutism. The majority of cases of deaf mutism are not congenital at all but due to severe middle ear infection in the first few years of life, which renders the child deaf before it has had time to learn to speak. In other cases where there is no history of infection you will find that a proportion, varying in different investigations from about 30 per cent to 10 per cent, are found to be the children of first cousin marriages.

These cases of deaf mutism are particularly interesting because owing to social circumstances one is able to trace the heredity of the condition. An albino is very unlikely indeed to marry an albino or an individual who is heterozygous for albinism. A Jones-Jones is unlikely to pick a Jones-Jones or a Smith-

Jones for a wife, and therefore albinism is very rarely transmitted from parent to offspring. In the case of deaf mutism, however, owing to the fact that they are segregated in special asylums, marriages between two deaf mutes are not uncommon, and it is quite frequent for all the progeny of such a marriage to be themselves deaf mutes. Very often, however, the children may be completely normal.

The reason will become clear if we change for a moment to a consideration of the situation in mice. There are four kinds of deaf mice whose heredity is very well understood. They have been chosen not because they are deaf but because this abnormality of theirs induces in them such habits as waltzing in a horizontal plane and shaking their heads in a vertical plane. The waltzers and shakers are invariably found to be stone deaf. There are four different genes giving deafness in mice, and they are all recessive. It is obvious that a union of two similar recessives will give nothing but recessives of that type. In other words, waltzing mice of a given type will be expected to breed true. On the other hand, a cross between waltzers of two different types, with different genes, will give normal mice. If there were asylums for the mouse population you would expect to find that the union of two waltzers would give waltzers, if, and only if, both parents differed from the normal in respect of the same gene. That is a rough approximation to what we see in men.

So far I have dealt with genes which occur in pairs, but there is a certain group of genes which behave differently. A woman has 24 equal pairs of chromosomes in the nuclei of her cells. A man has 23 equal pairs and one unequal pair, consisting of a large chromosome similar to one of the female chromosomes, called X, and a smaller chromosome called Y. All the ova of a woman contain an X chromosome, whereas the spermatozoa of a man fall into two classes; those carrying an X chromosome and producing females, and those carrying a Y and producing males. I state this dogmatically, because I regard it as proved, though I cannot go into the evidence here. A group of genes is carried by the X and Y chromosomes.

The most remarkable of these affected all the male descendants in the direct line of Mr. Lambert, an Englishman born in 1717. These men had a tough scaly hide except on the face, palms and soles. They moulted this once or twice a year. The abnormality was handed down from a father to all his sons. It never affected a daughter, nor any descendant of a daughter. In all 12 men, belonging to six generations, were affected. The last of them died in the nineteenth century. The inheritance was of the type expected if the character was due to a gene carried by the Y chromosome. Such genes are exceedingly rare in men, and we need not worry with them further. In the case of a character carried by the X chromosomes things are slightly more complicated.

To go back to our analogy with names, let us consider a community where every man has two names, one of which is the same. Thus in India all Sikhs are called Singh, which means Lion, after their more personal name. In our imaginary community a man is called Smith-Lyon or Jones-Lyon. He hands on the name Lyon to his sons, Smith or Jones to his daughters. The women, being less ferocious, have two ordinary names, such as Smith-Jones or Smith-Smith. They hand down one or other of these to a son or daughter. Thus a man inherits his single non-leonine surname from his mother and gives it to his daughters, but not to his sons. A woman inherits a name from each parent.

I will not deal with dominant characters carried by X chromosomes because only two such are known, and one of them is dealt with in a paper of mine now in the press. And my evidence for its existence may be judged inadequate.

Recessives in the X chromosomes are by no means uncommon, and a recessive character of this kind present in about 2½ per cent of the men of England is colour-blindness. In a pedigree such as that of Fig. 5 many of the males and no females are found to be suffering from it, and it is also found that the various men in the pedigree were related through women.

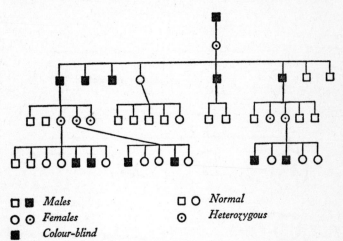

Males □ ■
Females ○ ⊙
Colour-blind ■

Normal □ ○
Heterozygous ⊙

FIG. 5.—*Pedigree of colour-blindness (protanopia or red blindness).*

In addition about half the women in the last generation were presumably heterozygous, but it is not known which.

The men never, or very rarely, hand down the condition to their own sons, but frequently hand it down to their daughters' sons. The reason for this peculiar type of inheritance becomes clear when we think in terms of chromosomes.

In Fig. 6 the X′ represents an X chromosome carrying the gene for colour-blindness. If a man has such an X chromosome he will be colour-blind because there

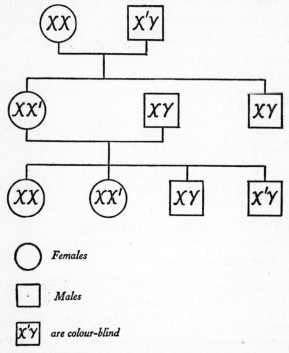

Fɪɢ. 6.—*Diagram illustrating inheritance of colour-blind-ness.*

is no dominant gene to mask its manifestation. If he marries a normal woman with two normal X chromosomes all of their children will receive a normal X chromosome from the mother; on the other hand the daughters will also receive an abnormal X from the father. They will usually be normal as regards vision, however, because the gene is recessive. The boys will receive a Y chromosome from the father, and will also be normal, and will not transmit the abnormality any further.

On the other hand, should a heterozygous woman marry a normal man she will transmit the abnormal X to one-half of her sons. Similarly, one-half of her daughters will be potential mothers of colour-blind sons.

If this particular type of inheritance were confined to colour-blindness it would not greatly concern us. Unfortunately, it also applies to a very much more serious disease, hæmophilia, in which the blood does not clot except after very great delay; and which has the effect that small injuries may be fatal, either from internal hæmorrhage or from loss of blood to the outside.

There are two forms of hæmophilia: a severe form such as exists in several descendants of Queen Victoria; and a milder form, such as exists in the pedigree of Fig. 7, where a large number of affected males were able to live long enough to marry. All spouses in this pedigree were normal. The original progenitor had two daughters, and each of these daughters had one or more hæmophilic sons; the same type of inheritance is demonstrated throughout the pedigree. It will be seen that the disease may be carried for no less than five generations in the female line before it shows up in a male. The gene, in other words, has

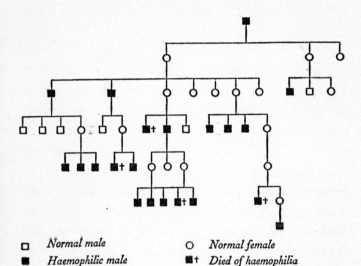

□	Normal male	○ Normal female
■	Haemophilic male	■† Died of haemophilia

Fig. 7.—*Pedigree of mild hæmophilia. All spouses normal. After Hay.*

been handed down through these five women, all heterozygous, who carried it, but their blood clotted normally because in addition to the abnormal gene they possessed normal genes which prevented it from showing its presence.

A number of other characters of varying degrees of severity are inherited in that manner. For example, there is an eye disease characterized by night-blindness,

67

myopia and nystagmus, an abnormality of the skin and teeth, and several forms of blindness.

We now leave the relatively secure ground in which the abnormal genotype is always distinguishable, to deal with those unfortunately much commoner cases in which a certain abnormality "runs in a family," but may skip a generation. In this case we can only say that the abnormal genotype does not always show itself in visible abnormality; or that if it so shows itself the abnormality is not always of at all a serious character.

Children are sometimes born with abnormally brittle bones. They may be broken before birth, and repeated fractures may lead to terrible deformity. The head is commonly of a peculiar shape, and deafness due to otosclerosis is not rare.

If we study the heredity in such cases as that we very often find the picture shown in Fig 8. In this pedigree the black circles do not represent brittleness of the bone; they represent a very minor abnormality; namely, a blue colour of the sclerotics of the eyes, the part which in normal individuals is white. The sclerotics are thinner than normal and the dark pigment inside shows through. On the other hand, vision is perfectly normal.

Among the children with blue sclerotics a number suffer from fractures and others suffer from deafness due to otosclerosis. Something like 40 per cent have one affection and about as many the other.

This is important, because the only character which obeys Mendel's law, the only certain manifestation of

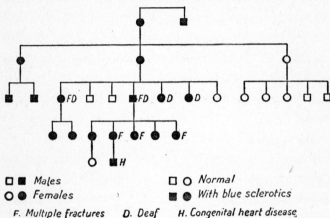

F. Multiple fractures   D. Deaf   H. Congenital heart disease

FIG. 8.—*Pedigree of blue sclerotics and associated abnormalities. All spouses normal. After Coulon.*

the abnormal gene, is the presence of blue sclerotics. For unknown causes, probably variations in nurture, a proportion of these people become brittle-boned.

That condition of affairs is by no means uncommon. For example, there is a type of hereditary jaundice in which on the whole the disease is handed down by a parent to about half of his or her offspring. But it quite frequently skips a generation in the pedigree (*cf.* Fig. 9). In such cases it is found that not only the jaundiced individuals but the individuals who have transmitted the disease without showing it have unduly fragile red blood corpuscles, which readily break up in a hypotonic solution. In the majority of such people a chill or an illness at some time in their lives breaks up the red blood corpuscles to such an extent

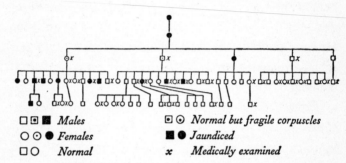

□ ◨ ▨	*Males*	◙ ☉	*Normal but fragile corpuscles*
○ ☉ ●	*Females*	▧ ●	*Jaundiced*
□ ○	*Normal*	*x*	*Medically examined*

FIG. 9.—*Pedigree of acholuric jaundice. All spouses normal.*
After Campbell and Warner.

that the blood pigment is converted into yellow bile pigment in such excessive quantities as to cause jaundice. But a man without fragile corpuscles may go throughout life without undergoing such a crisis.

In a similar condition found in coloured people in the United States the conditions are much more complicated. It is found that about 7 per cent of the coloured people have red blood corpuscles which assume the shape of a sickle or crescent, and the condition is due to a dominant gene. Among the people with sickle cells less than 1 per cent develop a painful and fatal form of anæmia complicated with jaundice.

Why, nobody knows, and until it was found that the peculiar type of blood cell in question was fairly common among American Negroes there was no real reason to believe that there was a hereditary element behind the fatal anæmia at all. It is in fact very rare to get two cases in one family. The sickle-shaped cells are handed down as a dominant hereditary character which renders people liable to die in this particular manner.

There are a great many other conditions which in the pedigrees skip a generation. For example, Huntington's chorea, a horrible disease of the nervous system which causes forced movements in its victims and usually drives them mad before they die, is inherited as a perfectly straightforward dominant, but the average age of onset is about 35. It is by no means infrequent in a pedigree to find that a parent has died at the age of 25 to 50 and transmitted the disease from one of his parents to one of his own children without living long enough to manifest it himself.

In the same way a number of diseases of the internal organs are in all probability due to dominant genes. Nevertheless, as these diseases are not invariably fatal and even where they are so are not invariably followed by post mortem examinations, it is impossible to say that they are invariably handed on and never skip a generation; even though we may regard this conclusion as probable.

I have not yet spoken of the question of how these abnormalities start. I have presented pedigrees of the kind which are found in the text-books which generally leave that important question out. I shall have to discuss it in the next chapter, and I shall also have to deal with the question of how far measures of negative eugenics such as sterilization would be efficient in stamping out the various abnormalities with which I have dealt; and whether any other measures might be more effective.

For those who are interested in the subject I particularly recommend perusal of two works: One is *The Treasury of Human Inheritance*, a vast series started by the late Professor Karl Pearson, containing an unrivalled collection of pedigrees. The other is Cockayne's book on *Inherited Abnormalities of the Human Skin and its Appendages*.

I have thought it best in this chapter to concentrate on the positive facts without drawing any moral from them. Nevertheless, I think it will be obvious that eugenic measures would have considerable value in some of these diseases and very much less in others.

CHAPTER III

The Origin of Hereditary Diseases by Mutation. The Possibilities of Negative Eugenics

SO FAR we have been discussing the inheritance of certain abnormalities. I have dealt with the kind of pedigrees that get into text-books. We have now got to consider how those abnormalities originate, or to take the thing back one step further, how the genes originate which are responsible for the abnormalities. In the early days of genetic research this question was not seriously raised, and there is still a tendency in certain quarters to neglect it. There is a belief that a condition transmitted to the descendants is always inherited from the ancestors, and a feeling of fatalism such as was expressed in the lines;

"With earth's first clay they did the last man knead,
And there of the last harvest sowed the seed,"

but actually when we are dealing with mice, flies, peas or something of which you can breed a reasonable number of generations in a human life-time we soon find that new genes do occasionally originate under conditions where close observation is possible.

73

The phenomenon of the origin of a new gene is called mutation. We are still to some extent in the dark as to exactly what happens. We do not know whether to regard the gene as an elementary organism which reproduces itself, or as part of the cell nucleus which is copied by some other part at each cell generation. What we can say is that the process by which the gene produces its like, whether we call it reproduction in the biological sense, or copying, is not an invariable process, that it breaks down with a certain small frequency, rarely more than once in a million cell divisions.

Let us take a case from man. Hæmophilia is due to an abnormal gene in the X chromosome. The abnormal gene in hæmophilia is a gene concerned in blood clotting which has become in some way inactive although it can reproduce itself; but it plays no further part in blood coagulation. Hæmophilics mostly die young, and on an average beget less than a quarter of the number of children produced by their normal brothers, and it is therefore clear that as a result of this frequently drastic selection there is a tendency for the hæmophilic gene to disappear. We may take it that roughly one-third of all such abnormal genes are in the X chromosomes of men and two-thirds in the X chromosomes of women, because a woman has two X chromosomes and a man only one. Therefore something less than one-third, perhaps one-quarter, of all hæmophilia genes in the population are wiped out in each generation. Unless then there were some source from which they can be replaced, the frequency of hæmophilia would be diminished by 25 per cent in each generation, and by a simple calculation from the existing frequency of hæmophilia it could be shown

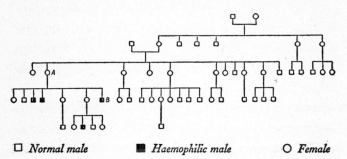

☐ *Normal male*   ■ *Haemophilic male*   ○ *Female*

FIG. 10.—*Pedigree illustrating the origin of hæmophilia by mutation. After Bell and Haldane.*

that at the time of the Norman conquest the entire male population of England must have suffered from hæmophilia. We have reason to believe that that is inaccurate. We therefore look for some evidence of the origin of new hæmophilia genes.

Such evidence is found in the pedigree of Fig 10. It is the pedigree of a family mostly living in northeast London, recently investigated by Dr. Julia Bell and myself. You will notice that all the hæmophilics are descended from one woman, marked A, who must

75

have been heterozygous for hæmophilia. Her father was not hæmophilic; she had five sisters none of whom produced a hæmophilic child; and she also had two normal brothers. There is further no record of hæmophilia among her mother's brothers' or among her mother's sisters' descendants. If we suppose that the gene for hæmophilia was handed down in the X chromosome from her grandmother through her mother we can calculate the chance that all these people should have been free from hæmophilia. We have a further bit of indication in this particular case. The man marked B is not only hæmophilic but colour-blind. The genes for hæmophilia and colour-blindness are found in the same chromosome and are generally handed on together, and in a given pedigree if one hæmophilic is colour-blind so will be the large majority of his hæmophilic relations. We know nothing of the vision in the other three hæmophilics, because all died of hæmophilia before there was any chance of testing their colour vision. We do know that a number of cousins of this hæmophilic are colour-blind, in other words that they have received at least a portion of the X chromosomes that in the man B also carried the gene for hæmophilia. Given this evidence we can calculate the probability that, on the hypothesis that the hæmophilia gene was handed down from A's grandmother, all these various males should have escaped; and it is less than one in five thousand. If that were an isolated pedigree we should have to accept it as a coincidence. But it is far from isolated. In our work on hæmophilia and colour-blindness Dr. Bell and I found altogether six pedigrees in which both of these conditions were associated, and in no less than three there was good evidence for the origin of hæmophilia by mutation. We assume that the hæmophilia gene arose in the one of the grandparents of the three affected brothers and was handed down to their mother only, who gave it to three of her four sons, and one of her three daughters. Such a conception would be in complete harmony with what we know takes place in many animals.

We found evidence of mutation in three of our six pedigrees, and it may well be asked why the evidence for mutation in the case of hæmophilia has on the whole been so slight in the past. The reason is a very simple one. The pedigrees to be found in the literature are generally pedigrees containing large numbers of hæmophilics. Pedigrees containing only a small number are not considered worthy of publication. The data, therefore, although perfectly correct, are considerably biassed in favour of pedigrees containing a large number of abnormals. It is only when one is interested in problems such as that of the origin of genes that one takes the trouble to investigate a large number of normal relatives. That was only possible in this instance through the cordial co-operation of our colour-blind hæmophilic subject B, who arranged for numbers of his relatives to turn up to a family party and have their colour vision tested. A simple calculation based on the frequency of hæmophilia in the population suggested that roughly speaking one normal gene in an X chromosome mutates to the hæmophilia gene in fifty thousand generations. That is the order of magnitude of the frequency. Certain American data suggest that it may be rather higher. Penrose found exactly similar results for a dominant gene causing a disease called epiloia which is responsible for tumours of various kinds, including tumours in the brain that give rise to mental defect. This condition is handed down as a dominant from parent to offspring, but as it causes considerable ill-health, due to disease of the brain, the heart, or the kidneys, it is rarely handed down for more than two or three generations, and is only kept in existence by mutation.

We may take it that where a gene lowers the fitness, assessed in the Darwinian sense on the basis of the average number of offspring left, there is a fairly close equilibrium between mutation and natural selection so long as we are dealing with dominant genes and sex-linked genes which come up to the surface frequently and are therefore exposed to the influences of natural selection. Recessive genes are far from being in equilibrium. It was earlier shown that about ½ per cent of the genes at a particular spot in the chromosomes of the Swedish population were recessive genes for juvenile amaurotic idiocy. The vast majority of these genes are present in heterozygotes and are therefore not subject to natural selection. Only when two happen to come together does natural selection work. The frequency with which a recessive gene appears in homozygotes depends on the amount of inbreeding in the population. The gene would be in equilibrium if as many were wiped out by selection as arose by mutation in each generation. Our ancestors a few thousand years ago were very much more inbred than we are. They lived in small communities, in isolated villages, or in small wandering tribes, and it is only comparatively lately that the amount of inbreeding has been very much diminished. When the amount of inbreeding is lessened the effect is to lower the frequency of homozygosis, and therefore idiocy becomes rare in the population, but the gene for it, not being subjected to such intense selection, becomes commoner. The gene frequency goes on increasing until a balance has been effected. A rough calculation gives an idea of the speed of that process. The time taken to reach the half-way stage towards equilibrium is something of the order of five thousand years or more. In other

79

words, we may look forward to a slow but sure increase in the frequency in our population of albinism, recessive types of blindness, and various types of idiocy, unless we can find some method of dealing with the problem. Fortunately the problem is not in the least urgent. The increase will take place at a very slow rate to be measured in thousands of years. We need not worry about the approaching degeneracy of the human species. We have at the present moment other and more urgent problems to consider.

So far we have dealt with abnormal physical characters whose heredity is fairly clear, but there are a great many cases where heredity is a fact but we cannot say much about it.

Table 5 summarizes the resemblance between relatives as regards cancer. The material is American. Of 953 sons whose fathers had died of cancer, fourteen had developed that disease. Of 95,300 men and boys of the same ages, but taken at random from the American population, eighty-five would, on the average, have developed it. The probability, as the result of chance, of finding fourteen individuals where 0.85 were expected is about 10^{-12} or one chance in a million million. On the whole, we see that the presence of cancer in a near relative increases a person's probability of dying of the same disease about tenfold. It does not of course make it certain, or even very probable.

80

TABLE 5

Inheritance of Cancer in Man

Cancerous Relative	Relatives Investigated	Number Investigated	Number Cancerous	Expected Cancerous on Chance Basis	Probability that Result is due to Chance
Father	Sons	953	14	0·85	10^{-12}
Father	Daughters	788	10	1·05	10^{-6}
Mother	Sons	565	12	1·19	10^{-8}
Mother	Daughters	504	27	1·59	10^{-23}
Sib[1]	Sibs	2,016	48	5·80	10^{-27}
Total		4,826	111	10·48	10^{-76}

After Little. N.B.—Little's figures in the last column are far larger than my own, that is to say the odds against chance are sometimes under a thousand to one. My calculation is based on the Poisson distribution. Thus if the number found is c and that expected on a chance basis is m, the probability is $\dfrac{m^c}{c!e^m}\left(1 + \dfrac{m}{c+1} + \dfrac{m^2}{(c+1)(c+2)} + \cdots\right)$. Little apparently assumed a normal or Laplacian distribution.

[1] I.e. brother or sister.

This is what would be expected if cancer is determined by a number of different genes interacting with the environment.

It might be thought that this tendency for resemblance between parent and offspring or between one sib and another was not due to biological heredity at all but to infection or similarity of environment. That may be the case to some extent, but it is quite certain that in mice and other animals the tendency to spontaneous cancer is largely inherited. Even in mice bred specially for susceptibility to cancer there is no question of cancer being determined entirely by heredity. This is shown by the fact that within a pure line the age at which a given animal will develop cancer is not at all certain. It may be at any time from four months to two years—a very advanced age for a mouse. We know something about the environmental conditions that will determine the age of onset of cancer of mice in a pure line. We know that some procedures will bring it on earlier and that others will put it back, nevertheless the hereditary element in its causation is strong.

We also know something of what happens in crosses. In cases of mammary cancer the inheritance is very strongly maternal. That is to say the children of a mother from the cancerous line are much more likely to get it than the children of a father from the cancerous line, although the father transmits to a certain extent. This shows that inheritance is not entirely determined by genes but partly by something outside the nucleus, possibly even to some extent by the milk.

We now turn to the practical applications of this knowledge, the suggestion that the race would be greatly improved if the unfit were sterilized. The existing law in Germany is as follows; "Anyone who is hereditarily ailing may be sterilized if in the experience of medical science it is with great probability to be expected that his progeny will suffer from severe bodily or mental hereditary disorders." The list of people regarded as hereditarily ailing includes sufferers from congenital feeble-mindedness, schizophrenia, manic depressive insanity, inherited epilepsy, Huntington's chorea, inherited blindness, inherited deafness, severe inherited physical malformation, and, rather curiously, severe alcoholism. In Germany a person can only be sterilized if he or she is actually abnormal in some way. Many eugenists would like to go a good deal further than that. There is before the country in England to-day a Bill which would permit the sterilization of "persons who are deemed likely to transmit a mental defect or a grave physical disability to subsequent generations" whether they themselves have or have not this grave mental or physical defect. The proposed British law goes some way beyond the

German law, although it does not contemplate compulsory sterilization in many cases where that would be legal in Germany. In both cases the absence of a quantitative definition is noteworthy. In Germany the expression "great probability" is used. In England it is stated that these people are "deemed likely" to transmit the mental defect. It seems to me that in a case of this kind some quantitative measure would be very desirable. As all motorists are considerably assisted in keeping the law by the knowledge that if they drive in towns at less than thirty miles an hour they have at least some claim not to be considered to be driving dangerously, so where a certain amount of emotion is involved and where entirely new legal principles are being brought forward it seems desirable to specify rather clearly what degree of probability of producing defective offspring is regarded as justifying sterilization. This is done in the American draft law quoted in Chapter I. I shall now examine the effects that would accrue from sterilization.

Where, as in the case of Figs. 1 and 2, the abnormal condition is due to a dominant gene that manifests itself in 100 per cent of cases and early in life, it is clear that sterilization would abolish all hereditary cases of the abnormality, leaving only those cases that are due to mutation, which in these instances would be a very small fraction of the total. Where, however, the disease is dangerous to life or alternatively lowers fertility, the proportion of all cases which is due to mutation is much larger. We see, therefore, that sterilization in the case of dominants would always be fairly effective, but that it would be least effective under those conditions where the disease is most serious.

In many cases it will be remembered that a dominant gene does not manifest itself in all instances. In Fig. 9, for example, the condition observed is a chronic type of jaundice transmitted by two individuals, both of whom were normal in the sense of never having had jaundice though both had an abnormality of the blood corpuscles that could be detected by experiment. If in this case only persons with jaundice had been sterilized the elimination would occur, but at a considerably slower rate than if all carriers of the disease could be detected. Indeed, if we considered it important to eliminate the jaundice, it might be desirable to sterilize a certain number of people who from the point of view of health were perfectly fit.

For instance, the eldest brother in the third generation was stoking a boiler at the age of sixty. We have already passed beyond the scope of the German law.

But the question arises whether even in the case of dominant disease this sterilization is the most desirable course. We have six possible alternatives if we consider it desirable to eliminate such a dominant gene;

85

(1) we might discourage the marriage of affected people and perhaps also of apparently normal carriers; (2) we might forbid it; (3) we might encourage continence either within marriage or outside it; (4) we might encourage birth-control; (5) we could try persuasion in the hope that these people would undergo voluntary sterilization; (6) we might have compulsory sterilization. Which of these alternatives is to be adopted is not, I think, a question for the biologist. It is a question of the relative value which is attached to various different goods, for example, to health on the one hand and liberty on the other. I personally regard compulsory sterilization as a piece of crude Americanism like the complete prohibition of alcoholic beverages. But I look to the common-sense of the American people to realize that here, as with prohibition, a mistake has been made. I would prefer myself to see a certain amount of government expenditure on propaganda among persons afflicted with hereditary blindness and other defects in favour of drastic family limitation by contraception or otherwise. It is perhaps characteristic that in the United States sterilization is legal while contraception is of very doubtful legality. I cannot help feeling that they have taken too rapid a step in practising compulsory sterilization.

To my mind the strongest objections to sterilization are two. The first of them is this, that whereas in the case of men it is a trivial operation, I will not say completely safe, but no more dangerous than the extraction of a tooth, which has been known to lead to death; in women it is a serious operation, perhaps about as dangerous as an operation for appendicitis in favourable circumstances. It is inevitable that if large numbers of women are sterilized, a certain proportion, I do not know how large—probably less than 1 per cent—will die as a result of the operation. Now it is a fundamental principle of English law that a person's life must not be endangered except in order to save him from some greater danger. That is one of the reasons why (in my opinion rightly) we forbid abortion at the mother's own request, and why we forbid a number of other procedures that endanger life and would inevitably lead to a certain loss of life. I am not at all convinced that this principle of the sanctity of human life may not be of somewhat more importance for the State in the long run than a reduction in the number of defectives of certain kinds. In Denmark quite recently one woman died as the result of this operation, and there have been reports in the newspapers of considerable numbers of such deaths in Germany. But I do not regard all news from Germany as being absolutely reliable. That is one objection. The second objection is that the demand for sterilization is a symptom of a certain state of mind which we shall have to examine later and which we may not find entirely admirable.

To continue with our biological approach. What would be the effect of sterilization on a sex-linked recessive condition such as hæmophilia?

In the quite unusual pedigree of Fig. 7 several of the males had children. In the average pedigree of hæmophilia transmission from a male is fairly rare. It is clear that if we wished to wipe out hæmophilia we should have to sterilize the females who are thought likely to transmit it. We could not sterilize the males because the operation would probably kill them, and nature sterilizes them already to a considerable extent by killing them off as children. We should have had, for example, to sterilize A in Fig. 10 as soon as she produced her first hæmophilic son. By so doing we should have prevented the birth of three hæmophilics and also of six normal children. We must consider what would be the results of such a policy. For example, the late Queen Victoria was heterozygous for hæmophilia, probably as the result of mutation. One of her four sons, Prince Leopold, was a hæmophilic. He was kept alive, married a German princess, and managed to transmit the disease to some of his descendants. Two of Queen Victoria's daughters were heterozygous and transmitted hæmophilia to the late royal families of Russia and Spain. There can be little question that the fact that both in Russia and Spain the heir apparent at the time of the revolution was a hæmophilic had a favourable effect on the revolution. We may, therefore, ask ourselves first whether, had a law as to sterilization been in force during the nineteenth century, it would have been applied to the reigning monarch and her daughters; and secondly, if so, what would have been the effects on European history.

When we come to ordinary recessives we can say at once what the effects of sterilization would be. Except in cases where, as with deaf mutism, there is a tendency for recessives to inter-breed, there would be no noticeable effect in less than thirty or forty generations.

But that is not to say that eugenic measures are impossible. The eugenic measures available in such a case are of an entirely different type. They include the prohibition or discouragement of cousin marriages. For example, were the marriage of first cousins forbidden this would cut down the incidence of juvenile amaurotic idiocy by some 15 per cent, congenital deaf mutism by about 25 per cent, xeroderma pigmentosum (a fatal skin disease) by nearly 50 per cent, and so on. It is interesting that the only body in this country which advocates that particular form of eugenics is the Roman Catholic Church, which is opposed to other eugenic activities. It is unfortunate that it is possible on payment of a suitable sum for Catholics to obtain a dispensation to marry their first cousins. However, so far as it goes the Catholic discouragement of cousin marriage must be regarded as a eugenic measure. Whether such marriages should be forbidden is a quantitative question. It must be said at once that the large majority of the children arising from the marriage of first cousins are perfectly normal, that if you marry your first cousin you will probably not produce a defective child. It is equally true, however, that you are much more likely to produce a defective child of certain kinds than if you marry an unrelated person.

The relative value to be attached to health on the one side and liberty on the other is not a question for the biologist as such. All that he can do is to put the pros and cons before ordinary people who are perhaps less biassed in some ways than himself, and leave the matter to them to decide.

The second eugenic measure that is thinkable in cases of recessive complaints is the dissolution, voluntary or compulsory, of the marriage that has produced one recessive child, and a third possible measure is to sterilize one partner of such a marriage.

A question which immediately arises is the relative frequency of recessive defects as compared with those that are dominant or sex-linked. That is a very difficult question to answer. In my opinion the recessive defects are considerably rarer, and Levit, after an extensive study of Russian populations, holds the same view. So far they appear to be mainly confined to the skin, eyes, and nervous system. But it seems to me that an investigation of the relative frequency of these different kinds of defects is absolutely essential before we can hope for any sane eugenic policy based on a full knowledge of the facts. Students of genetics in search of first-hand information are surprised at the rarity in the population of certain conditions which occur pretty frequently in text-books because they are themselves striking and are inherited in a striking and simple manner. Other congenital abnormalities which are unfortunately commoner are not always inherited in so simple a manner, and therefore less emphasis is laid upon them either by propagandists or by writers of text-books.

We now come to the question of mental defect. There is a demand that all mental defectives should be sterilized, or alternatively should be given the opportunity of being sterilized. The statement is made that they are very prolific. The more serious kinds of mental defectives, idiots and imbeciles, are not prolific; the feeble-minded are moderately so, though here there is some exaggeration, because a considerable proportion are in institutions where procreation is not possible. We saw that there are many different types of causation of mental defect, and if we try to discuss a comprehensive programme like that of sterilization we have to lump them all together and discuss what would be the effects of that measure.

Fortunately there are some very good data available concerning Birmingham in the *Report of the Special Schools After-care Sub-Committee* for 1933. The report was on three hundred and forty-five schoolchildren whose fathers or mothers, or in four cases both parents, had themselves attended a special school for mental defectives. Of these three hundred and forty-five only twenty-five or 7.5 per cent were at special schools or reported as for special school examination. If therefore the parents had been sterilized we should have got rid of these twenty-five defective children and also of three hundred and twenty children who at any rate were not mental defectives. As a matter of fact, of the non-defective children 18.5 per cent were backward, the majority were normal, while 3 per cent were actually above the average, a few even going to secondary schools. There is no doubt that in educational achievement they were on the whole below the average. Of the nineteen certified mental defectives no fewer than four were the children of one particular woman. Among these Birmingham mental defectives there are therefore very great differences as regards the capacity for transmitting mental defect to offspring. If we had a causal analysis of mental defect there would probably be a good case for preventing that particular woman from having progeny. Thirteen children had both parents defective. Of these thirteen only one was pronounced defective in 1933 and another one may possibly be registered by now.

This is of some interest, because it is often stated that the children of two defectives are all defective. This would be true if all mental defect were due to the same recessive gene, in which case, incidentally, it would rather rarely be inherited from a parent. It is also of interest because it suggests how much, or how little, we can salvage from the Jeffersonian doctrine of equality. I think that the following proposition would be accepted by most biologists: "It is never possible, from a knowledge of a person's parents, to predict with certainly that he or she will be either a more adequate or a less adequate member of society than the majority." In a very few cases, it is true, we can predict with certainty that a given unborn child, if legitimate, will have a certain physical defect. Thus two albinos probably always produce albino children. But our knowledge of the heredity of psychological characters, desirable or otherwise, is insufficient to make predictions of this kind. We can, of course, make statistical predictions. But we do not, in my opinion, know enough to accord rights to any individual, or to deprive him or her of any rights, on the basis of ancestry only. We shall see that this appears also to be true when we consider the hereditary differences between human races.

It will be remembered that about 7½ per cent of the children of the Birmingham defectives were themselves defective. One finds much the same proportion if one examines the parents of mentally defective children. For example, in East Suffolk Grundy followed up the parentage of one hundred and fifty-eight mentally defective children. Of these 6 per cent had parents mentally defective in the sense of being unable to earn their living, and 25 per cent had parents who were mentally defective from the educational point of view in the sense of not being able to profit by education, although they were able to earn their living. We may take it then that the sterilization of all mental defectives would probably cut down the supply of mental defectives in the next generation by something of the order of 10 per cent—some people put it as low as 5 per cent, others as high as 30 per cent. A very great difficulty arises because mental deficiency is a legal and not a biological conception. I am referring to what would happen if the people who are legally certified as defectives were prevented completely from having progeny. Although some mentally defective children would not be born, a considerable proportion, perhaps ten times as many, normal children would not come into the world.

After such data it is somewhat surprising to read the statements made by propagandists for eugenics. For example, Dr. R. B. Cattell, a Research Fellow with a grant from the Eugenics Society, writes in the *Eugenics Review* (28, p. 190): "About 75 per cent of the children of the feeble-minded are themselves feeble-minded, and the remainder are not far above the border-line." [1] If that is correct it can only be explained on the assumption that the feeble-minded of Birmingham are more desirable as parents than the feeble-minded elsewhere. I am, however, inclined to think that Birmingham is not unique in this particular respect. That this view is generally taken appears from the statement of Dr. Blacker, Secretary of the Eugenics Society, who writes: "It would therefore be idle to expect appreciable results from legalizing a measure of voluntary sterilization limited to mental defectives." He takes the view that such a measure should extend not only to mental defectives but to people who are definitely somewhat below the average as regards intelligence. With mental defects as with

[1] Dr. Cattell has since explained that he was referring to children both of whose parents were mental defectives. Even here Birmingham seems to be very lucky!

physical defects, if you once deem it desirable to sterilize I think it is a little difficult to know where you are going to stop.

Certain statistics suggest that an entirely different set of eugenic measures might be of great value in reducing the incidence of mental defect. Russell investigated the very bright and the very dull children in the rural elementary schools of an eastern county of England. He found that of sixty-three children with an intelligence quotient below 80 per cent, no less than twenty-five were the offspring of parents born in the same village. On the other hand, of thirty children awarded free places in secondary schools, only two were the children of parents born in the same village.

It might be thought that the parents of the backward children were the village idiots who had never had a chance of leaving their homes. This was not the case. One had been a soldier, another was a carrier, and so on. It is at least arguable that the backwardness of the children was largely due to inbreeding, which presumably caused recessive genes to appear in the homozygous condition. Russell's work has never been followed up. If it is confirmed, it is likely that the introduction of motor omnibuses into our rural areas will prove to be a eugenic measure quite as valuable as sterilization.

Among the conditions which are regarded as fitting a person for sterilization in Germany is schizophrenia (a form of insanity). About 8 to 10 per cent of the children of sufferers from schizophrenia are themselves schizophrenics. The proportion is very small, and is about what we find for mental defectives. Only 30 per cent of these children are born after the first admission of their parents to hospital for the disease. Sterilization of all schizophrenics in Germany is therefore not likely to reduce the numbers in the next generation by more than 3 per cent. Thus to prevent the birth of one child destined to schizophrenia in the next generation we must sterilize about sixteen schizophrenics and prevent the birth of ten normal children. It is highly questionable whether the bad effects of this policy do not outweigh the good effects, at least from the point of view of Hitler, who wishes the German population to increase.

I must now turn for a moment to the words "fitness" and "unfitness," words to which I object very strongly when used as eugenists often use them. The word fitness was used by Darwin in a perfectly intelligible sense to refer to individuals of such a constitution that they are likely to propagate themselves in larger numbers than their fellows, either as a result of being better adapted to their environment or more fertile, or both. It should be pointed out, however, that a lessened fertility may lead to a larger survival of offspring. A prize-winning hen laying three hundred eggs a year, if she had to fend for herself, would probably not rear so many young as an ordinary barnyard fowl. Fertility may or may not contribute to fitness. We also use the word fitness to denote the character of a good Rugby football player and so on, and there is a tendency to think that if a person is fit in one direction he will be so in another.

As a matter of fact this is not the case, unless we take the view that the unskilled workers of this country are fitter, not only than the capitalists, but than the skilled workers. For the unskilled are not only more prolific, but more of their children reach maturity, and they are therefore fitter in the Darwinian sense. It is worth remembering that there is no such thing as general fitness even in animals. In mice resistance to disease is an important element in fitness. Lines of mice which are resistant to mouse typhoid generally tend to be particularly susceptible to virus diseases such as "louping ill," while those which go down readily with typhoid stand up to virus diseases. We do not know whether this is a general law, but it may well be so.

In any case, when we use the word fit, we must ask "fit for what?" And that brings us up against the whole question of social ideals. A biologist may be pardoned if, in a biological discussion, he prefers to

98

use the word fitness in the same sense as Darwin used it. If it is used in that sense, we find that in many cases the eugenists are demanding the sterilization of the fit. This is not a criticism of the eugenic programme. Man should not follow nature blindly. He should, and does, interfere with natural processes, including natural selection. But it is a criticism of eugenical terminology.

It is worth while examining very briefly the psychology of the demand for sterilization and the belief that it would have important social consequences. In the first place I think that it depends to a considerable extent on one's emotional reaction to mental defectives. Many people regard them with horror; personally, I must confess to a certain liking for them; and this is shared, curiously enough, by many people who work with them, teaching and supervising them in special colonies. Thus Sutherland says: "The smiling face of the Mongolian imbecile suggests the possession of a secret source of joy." I am prepared to believe that one's emotional reaction to defectives biasses one's attitude on the question of sterilization. It is beyond my purpose, however, to go into the psychological causes which may lead one to a feeling of horror, and another to a feeling of affection.

A second determining motive is the feeling of fatalism that some people have with regard to congenital defects. It was Oscar Wilde who said of heredity: "It

99

is the last of the fates, and the most terrible. It is the only one of the gods whose real name we know." To regard anything which we can understand as a fate appears to me extremely unscientific. Hegel made the very profound remark that freedom is the recognition of necessity. When we can really understand the nature of a phenomenon, we are some way on the path towards controlling or circumventing it.

At an earlier stage in this discussion we considered a pedigree of blindness, due to retinitis pigmentosa (Fig. 3). It is claimed that this disease can be controlled by injections of Vitamin A, which plays an important part in the chemical processes that go on in the retina and lead to vision. Whether that claim is correct I do not know. It is of importance to realize, however, that three or four hundred years ago, before spectacles were invented, extreme short sight, which is often strongly inherited, would have been regarded as partial blindness; it is now only the justification for buying suitable spectacles. An extreme eugenic attitude which encourages a fatalistic approach to maladies that have an hereditary element is to be deplored. There is another unconscious motive behind the demand for sterilization. Psychoanalysts claim that internal contradictions in the human soul are unconsciously projected as hatred, and that in a society where this must be normally repressed it is readily directed at objects discredited for some reason. I should be the last to flatter myself with the thought that my own motives for opposing sterilization are wholly rational, for it is important to realize that one cannot avoid irrational motives.

Finally, there is a certain amount of motivation in its favour by the class struggle. In the past we have had the feudal theory of noble blood and the monarchist theory of divine right. These are now discredited, but many able men believed in them in the past. Their modern equivalent is, I suppose, the doctrine of the innate superiority of the children of the well-to-do.

Before I turn to that question I wish to say a few more words about the Voluntary Sterilization Bill which is now before the country; the text of it will be found in the *Eugenics Review* of July 1935. It permits of the sterilization of four classes: (1) mental defectives; (2) those who have suffered mental disorder in the past; (3) those who have grave physical disability which is likely to be inherited; and (4) those who are deemed likely to transmit a mental defect or grave physical disability to subsequent generations. It is true that a medical certificate is needed before sterilization can be carried out; nevertheless, I confess that I consider the title of the Bill to be a dubious piece of terminology, since it describes sterilization in the case of mental defectives as voluntary. Dr. Penrose [1] quotes a letter from Dr. Turner, who writes: "I venture to say I should not be fitted to hold my present office of medical superintendent of an institute for the care of mentally defectives if I could not induce practically every one of my patients to be operated on or to refuse an operation just as I myself might wish." Even where sterilization is voluntary we need a guarantee that it should be truly so. In the United States this operation is sometimes performed under circumstances where the will of the person sterilized has been influenced to a considerable degree. Let us take the case of John Hill. I choose this case because it is given in Laughlin's *Eugenical Sterilization in the United States*, a book which is written in defence of that operation. I cannot, therefore, be accused of perverting the truth if I quote it verbatim. Other cases more favourable to my own views have perhaps been reported by opponents of eugenics, who may have suppressed certain facts. Hill was tried by Judge G. B. Holden of the Superior Court, County of Yakima, Washington, U.S.A. Here are Judge Holden's words:

"The case in question is that of State of Washington v. John Hill, upon whom I suspended judgment and suggested an operation for the prevention of

[1] *Mental Defect*, p. 170.

procreation. This, however, was merely a suggestion, and not a part of the judgment in the case.

"On January 30, 1922, John Hill pleaded guilty to the crime of grand larceny. The theft was of a number of hams, which he took by stealth because of his impoverished condition; their value, however, being more than $25.00, he was guilty of grand larceny and subject, under our indeterminate sentence law, to not less than six months, nor more than fifteen years, imprisonment in the state penitentiary, which was the judgment of the court and the judgment was suspended during good behaviour. The facts of the case, which led to the suggestion that he submit to a voluntary operation for the prevention of procreation, and to which suggestion he assented after the details of the operation (vasectomy) and its results were explained to him, are as follows:

"Hill is a Russian beet sugar laborer, with a wife, and five children all under the age of eleven years. He is robust physically, about forty years of age, and his wife some years his junior. Hill, his wife and five children are all mentally subnormal, even for their situation in life. For many months the children have been half-starved and half-clothed. It was apparent that he could not provide them with the common necessities of life, to say nothing of giving them any sort of advantages in the world by way of education or other preparation to battle for themselves. He was forced to steal to prevent them from starvation, or to apply for public aid. The case was brought to the attention of the authorities through the discovery of the theft of the hams, since which time he and his family are partially dependent upon public charity, and without the addition of more children to the family will undoubtedly continue to be more or less a public charge; with more children the extent of demand for public charity will be increased. Under these conditions the operation was suggested to him, and after explanation, as before stated, he consented." [1]

We are not told whether the consent would have been obtained so readily had the suggestion been made by a man who had not the power to send Hill to prison for fifteen years. Nor is it clear what tests were employed to detect the mental subnormality of the Hill family. Some quite intelligent people do not appear at their best in a criminal court. The type of evidence on which Judge Holden based his eugenical activities may be inferred from his statement concerning Chris McCauley, a burglar whom he sentenced to compulsory sterilization.

"This man, about thirty-five years of age, is subnormal mentally and has every appearance and indication of immorality. He has a strain of Negro blood in

---

[1] *Eugenical Sterilization in the United States*, p. 92.

his veins, and has a disgusting and lustful appearance."

It is, I think, clear that Hill would not have been sterilized had he possessed an independent income. And it is unlikely that McCauley would have been, had his complexion been lighter and his appearance more in conformity with Judge Holden's aesthetic standards. In my own judgement at least one well-known cinema "star" "has a disgusting and lustful appearance," but I claim no scientific basis for this opinion, nor do I suggest that it be used as a basis for eugenical sterilization. In view of the judgements quoted it is of interest that British eugenists often state that in America sterilization is carried out with a complete disregard for class or race distinctions.

Should sterilization ever become compulsory, or even legal in England, it is of the utmost importance that orders or even suggestions for its application should be made quite impartially. At the present time nearly all cases of lunacy or gross idiocy are probably reported in all social classes. However, there is reason to think that mild mental defect is much less frequently certified among the rich than the poor. A well-to-do family can afford to keep a "backward boy" or "a girl who was no good at school." A poor family cannot. Sterilization of all certified defectives would thus in our society be a class measure.

Incidentally it is commonly stated that the greater

frequency of feeble-minded (though not of idiot) children among the poor is a sign of their innate *Minderwertigkeit*. It may be so, but the above explanation is equally plausible. Only a very careful investigation, by investigators legally entitled to investigate the skeletons in the cupboards of the rich, could decide the question. And there are many more important things to investigate.

I have no great confidence in the wisdom of our judges where human reproduction is concerned. Forty years ago they took every opportunity to fulminate against birth control. Now some of them go to the other extreme and use their positions to blame parents of large families for their fertility if they are unable, at existing wage rates, to afford them a suitable upbringing. Without going so far as Judge Holden, they might allow economic factors to influence them when sterilization was concerned. It is true that the Bill to which I have referred requires the consent of two doctors, one specially appointed by the Ministry of Health. But doctors are not in general taught human genetics. A medical student who has attended three lectures on the entire subject of genetics is unusually well informed. It is true that in 1938 it is proposed to include genetics in the medical syllabus. So by 1950 or so a fair proportion of doctors should be informed on this topic. Whether they will be so is another question. For very few people in England to-day are qualified to teach genetics, either human or animal, and so far from any attempt being made to remedy this lack, the teaching of genetics in London is at the present moment being drastically cut down.

It is not only from the legal authorities that pressure is to be feared. A well-known employer in Ontario during the recent depression offered to pay for the sterilization of a number of his workmen whom he regarded as mentally dull, and several consented. During a time of unemployment it is generally advisable to comply with the suggestions of one's employer. Should a civil commotion ever occur in Ontario during which the employer in question is deprived of the opportunity for reproduction, this will, of course, be an unmentionable atrocity. But there is apparently nothing wrong with a measure which tends to diminish the population of a Dominion which could admittedly support a very much larger number than it now contains.

Finally a further problem arises. If mental defectives are to be kept permanently in an institution, there is clearly no need to sterilize them. If they are let out after sterilization, several alternatives lie before them. They may obtain employment. An acquaintance informed me that he preferred feeble-minded men to look after his pigs. Penrose [1] writes as follows: "A striking feature of defectives of imbecile and lower (i.e. less serious) grades is their apparent incapacity for being bored with an occupation; and provided some simple manipulation can be taught, the defective is perfectly happy in continuing the same simple manipulation for days and years without any change. This fact makes possible methods of dealing with patients who might otherwise be difficult to employ. In a regular, even if very monotonous employment they learn to be useful and worthy people."

If this statement is true, it suggests that mental defect is to a large extent a social rather than a biological problem. In a society where there was work for all, and vocational selection, places would be found for many, perhaps the majority of people who are now regarded as feeble-minded. The large increase in recent years of the number of people certified as feeble-minded may turn out to be a result of the increasing difficulty in finding regular employment rather than of any rise in the number of people falling below a certain grade of intelligence. In fact it may be a social and economic rather than a biological phenomenon.

I am of the opinion that a man who can look after pigs or do any other steady work has a value to so-

[1] *Mental Defect*, p. 164.

ciety, and that we have no right whatever to prevent him from reproducing his like.

The discharged defective may merely go to swell the number of unemployed, in which case he or she would almost certainly have been happier in an institution, and probably less trouble to society. But there are two much less satisfactory possibilities.

A number of the sterilized girls in California have married after discharge, and in many cases their husbands express themselves as satisfied. If they are genuinely feeble-minded, this does not speak very highly for the husbands. To the majority of people marriage means a great deal more than legalized sexual intercourse without the possibility of procreation, and any course of action which reduces it to that level appears to me to be at least as antisocial as one which allows an occasional defective to be born.

Finally, the discharged defective may be boarded out with a family who look after him or her, and give him or her employment at a wage below the standard level. If such persons are not closely supervised by Government agencies, they degenerate into slaves. In any case they tend to lower wages and displace more highly-paid workers. The problem of discharged defectives is already causing some disquiet from this point of view in Denmark.

In a society where there was work for all much

might be said for a system under which even mental defectives contributed to some extent to the common good by working so far as was possible. But our society is not of that type. There can be no argument for releasing genuine mental defectives except the economic argument that it is expensive to look after them. This expense is considerably exaggerated. It costs about £1 per week to keep a defective in an institution. An unemployed man costs from 10s. to 16s. per week, and more if he is married. There are about 300,000 mental defectives in England and Wales. The majority are at large. The cost of segregating them all would be about the same as that of an additional 100,000 unemployed, and as they would not marry wives or produce children many of whom require unemployment benefit, there might be an actual saving after a few years. There would also be the same biological advantages, whether large or small, as would accrue from sterilization.

We are constantly asked why sound people should be taxed to support the unsound. The answer to this is not a matter of biology. In most human societies it is regarded as a duty to help our weak or unfortunate fellows. This may be a fallacy.[1] I do not think that it

---

[1] E.g., Miss Peterkin, a firm American believer in eugenics, writes in *Living Philosophies* (p. 200): "I do not believe that the care and pity given by the strong to the weak have helped civilization." Similar views have been expressed in England and Germany.

is, but I clearly cannot argue the matter here. It is of course a much harder question just how much effort should be devoted to this end. But it is at least arguable that the proposal to turn out a number of mental defectives into the bitter economic struggles of modern life, provided only that they cannot reproduce, is a step morally backwards, and an abandonment of one of the forms of behaviour which distinguish man from most other animals.

Others hold that this care and pity are an essential part of civilization, and worth while for their own sakes. If they tend to make civilization self-destructive by fostering the spread of undesirable genes, I suggest that the remedy can be found in the exercise of still more care and pity by segregating mental defectives under humane conditions.

CHAPTER IV

Differential Fertility and Positive Eugenics

WE MUST now take up the question of the different rates of increase or decrease in different nations and social classes. We must consider the causes and effects of these differences, and the methods which have been adopted or suggested for controlling them.

If we wish to compare the rates at which two human groups are increasing or decreasing, there are several possible comparisons. We may compare the birth-rates. It is obvious that a population with a high birth-rate is not necessarily increasing in numbers. The birth-rate in China is certainly higher than in most European countries. So is the death-rate. But we do not know whether the population is increasing or not. We may measure the excess of births over deaths. That is valuable. It gives the rate at which a population is now increasing or decreasing, but it affords little guidance as to the future. In order to predict the distance that a motor car will go in the next minute we should know not only the speed but whether the driver has his foot on the accelerator. To predict the trend of the population in the future we need statistics of a rather special character.

Consider one hundred thousand English baby girls born in 1920, and therefore now just coming to the age when a few of them will be producing babies themselves. Somewhat over ninety thousand of those girls are still alive, and most of them will survive until the age of fifty, after which they will not bear any more children. Now the average fertility of women at any given age is known. Fig. 11 shows some American figures. It will be seen that on the average one thousand women between fifteen and nineteen bear something like fifty children per year; between twenty and thirty they bear somewhere between one hundred and one hundred and fifty children per year, and after that, their fertility diminishes. It will also be noticed that the fertility at any given age was diminishing between the years 1920 and 1929. If we take one hundred thousand baby girls and allow for the fact that a certain number of them are dead at any given age, we then ask how many children they will produce, given the standard of fertility of women at each particular age. Clearly, if they have more than one hundred thousand girl babies in the course of their lives, then the population will tend to increase; if they have less than one hundred thousand, it will tend to decrease. The ratio of the number of daughters to the number of mothers is called the net reproductive index. We can make our calculation a little better if we allow for the trend of fertility at different ages. As the fertility of women between twenty-five and thirty is falling [1] in England it is reasonable to suppose that ten years hence it will be somewhat lower than it is now. How much, I cannot say. Nevertheless, the net reproductive index is usually calculated on the basis of the existing fertility at any particular age. The net reproductive index fell below unity in England, in the United States, in Germany, and in a large number of other European countries in the years between 1920 and 1930, and to-day it is well below three-quarters in England. The births in England, however, are slightly in excess of the deaths, and it is generally believed this country will attain its maximum population about 1940. After that it will gradually decline. The importance of the net reproductive index is that it gives us an indication of the future trend of the population. Although the net reproductive index is low in this country, the birth-rate is still relatively high because there is a very considerable proportion of women between the ages of fifteen and forty-five—the ages between which they bear children. That proportion will decrease in the future and the birth-rate will as-

[1] There was a slight, but very slight, rise of the birth-rate in 1936.

114

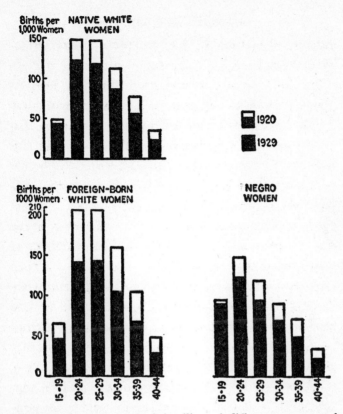

FIG. 11.—*Average annual fertility of different groups of American women as in 1920 and 1929. It will be seen, for example, that 1,000 American-born white women between the ages of 20 and 24 inclusive produced on the average about 148 children per year in 1920, about 125 in 1929. The high fertility of foreign-born women is largely due to the fact that most of the immigrant women of marriageable age are brought in as wives. It has not therefore the alarming significance which some American publicists have attributed to it. (From Lorimer and Osborn's* Dynamics of Population, *p. 42, Macmillan, 1934. This book is a mine of invaluable information, very fairly presented, although I do not personally agree with all the authors' conclusions.)*

sume the low level to be expected from the low reproductive index.

The reproductive index is below unity in most civilized countries. In Italy it is falling towards unity, and
is now close to it despite the efforts (counterbalanced
in recent years by the absence of many men in Ethiopia
and Spain) of the Government to keep up the birth-
rate. In Germany it fell considerably below unity, and
then rose sharply after the advent to power of the
National Socialist party in 1933. It is thought that this
is partly due to the system of marriage loans given to
young couples on marriage, and to propaganda. Glass,
in his recent book *The Struggle for Population*, takes
the view that the rise in the German birth-rate was
due to a considerable extent to the strict enforcement
of the laws against abortion. However that may be,
after an initial spurt the reproductive index is falling
towards unity, although it is still considerably above
it. The only important countries for which statistics
are available in which the net reproductive index is
steady and well above unity are the Soviet Union and
Japan. In the Soviet Union there was in recent years
a slight fall, but lately it apparently rose again when
abortion was made illegal. It is clear that many years
must elapse before the net reproductive index of the
Soviet Union falls to unity, and even after that, perhaps ten or twelve years before the birth-rate falls as

low as the death-rate and the population becomes stationary. We must look forward to an increase of the population of the Soviet Union for at least another fifty years; and if it continues to increase while the populations of various capitalist countries, including the United States and the British Dominions, are stationary or falling, that biological fact will presumably have a considerable bearing on the success or otherwise of Communism on our planet as a whole.

With regard to the cause of this mothers' strike, as it may be called, we are much in the dark. The best discussion known to me is that of Enid Charles in *The Twilight of Parenthood*. The use of contraceptives is a cause, but not the only cause. That is shown by one simple fact. In Ireland the fall in the birth-rate started before it had started in any other country. Ireland is a predominantly Catholic country in which contraceptives are little used. It started there apparently owing to economic causes; and apart from contraception, one cause of the falling birth-rate is unquestionably the increasing postponement of marriage. The natural fertility of women is probably at its maximum somewhere about seventeen or eighteen, and if they postpone marriage until they are getting on for thirty they are less likely to produce a given number of children. Abortion may be an important cause, especially in some Continental countries; and voluntary abstinence must be considered. It is a striking fact that among the Catholic peasants of Bavaria the drop in the birth-rate in the years between 1910 and 1920 was greater than among the Protestants and Jews living in the same country, although their absolute birth-rate was higher, both before and after. In this fall birth control by contraceptive appliances can have played no appreciable part.

There may also be a drop in fertility due to unknown causes. It may be that certain physiological features in our environment are less favourable to fertility than in that of our parents. This question is discussed by Charles. Whatever our answer is I would point out that it is a question that is extraordinarily important for the future history of the world.

Within most countries the fertility is sharply graded as between the different economic classes.

Table 6 shows some American data; and the British data are roughly similar. The fertility of American farm labourers was considerably more than double that of the professional classes. Fertility was higher among the poor than the rich, and higher in rural than in urban populations. These differences are fairly typical of the conditions in Europe, including England, although in England the agricultural workers are less fertile than the average, and the miners seem to be the most fertile large group. In a few towns such as

Stockholm the fertility of the poor is as low as that of the rich.

The situation shown in Table 6 is fairly universal.

TABLE 6

Relation Between Fertility and Social Status in America

Class	Total Children per 100 Wives aged 40–44 years in 1910
Professional	211
Business and clerical ..	224
Skilled manual workers..	277
Unskilled manual workers	334
Farm owners	376
Farm renters	467
Farm labourers ..	471

After Sydenstricker and Notestein.

These figures were compiled on behalf of the Milbank Memorial Fund on a sample of 100,000 American-born white couples in the northern United States. The marriage rates in the different classes do not differ greatly.

It means that economic success is the opposite to biological success. Darwin in describing Natural Selection observed that biological success is measured by the number of offspring left when allowance has been made for the number of deaths. The mortalities in

119

different classes under the age of forty-five or so are nowadays so nearly equal that they do not in any way compensate for the differences in fertility as they would have done one hundred or one hundred and fifty years ago, when infantile mortalities were over 50 per cent in English towns, as they are among the poor of many towns of India. The principle that biological success and economic or social success are to some extent contradictory is not a new one. So far as I know it was first enunciated nineteen centuries ago by Jesus Christ, who said: "Blessed are the meek, for they shall inherit the earth." Unlike the other beatitudes, this one can be verified, and, to a large extent, it is true. It was a remarkable statement at the time it was made. The Romans had just conquered the Mediterranean basin; they appeared to be inheriting the earth; and they were not meek. The only obvious alternative to the Romans might have seemed to be the Jewish Nationalists, who were not meek either. It seems probable that the people who actually left the largest number of descendants in the Mediterranean basin were not the Roman conquerors or the Jewish fanatics but the ordinary people who do not appear to any extent on the stage of history as usually written.

One might go so far as to say that a large part of the eugenic movement is a passionate protest against the hard fact that the meek do inherit the earth. The contradiction between social and biological success is not peculiar to our own civilization. In the Middle Ages I suppose that the qualities which were most admired were holiness in the clerics and valour in the nobility. The clerics were celibate, and the English nobility wiped itself out at Barnet, Tewkesbury, and Towton. It would be extremely interesting to know whether the same contradiction holds in the U.S.S.R. It would seem that many of the Communist leaders who are most admired there have small families or, like Lenin, have no families at all. It would be of interest to discover whether the rank and file of the party are less fertile than the average of the population.

We may take it as a general rule, whether we like it or not, that in a great many civilized societies those types which are regarded in the particular society in question as admirable are less fertile than the general run of the population; these societies would seem to be biologically unstable.

What effects can we attribute to this differential birth-rate? There is no question that if we take scholastic achievement as judged by examination or intelligence tests, which are supposed to be independent of the precise kind of school teaching, the children of the well-to-do fare very much better. In Northumberland according to Duff and Thompson among one thousand seven hundred and twenty-two children of parents classed as brain-workers fifty-nine, or 3.4 per cent, attained grade A in an intelligence test, whilst eight attained grade A +. Among ten thousand eight hundred and forty-eight children of hand-workers only seventy-nine, or 0.73 per cent, reached Grade A, and eight reached grade A +. Examples of the same kind could be multiplied. Exactly the same is found in areas of the United States where the school system is more uniform than in England.

The children of the professional classes only overlap those of unskilled workers and small farmers to the extent of about one-quarter in respect of achievement on intelligence tests (Fig. 12). The difference between fertility in different social classes is likewise clear; and we may also deal with the matter directly and consider the correlation between family size and intelligence. Until one gets down to definitely defective parents of whom a considerable number are segregated, one finds large family size associated with inferior intelligence. The family size of parents of inferior intelligence is nearly double that of those of a higher mental standard. We can say that if the intelligence quotient is determined by heredity, its average in the population will diminish. We must therefore ask how far the intelligence quotient is affected by nurture, and what elements affect it.

Professor Newman of Chicago studied a pair of monozygotic twin girls; they were born in Chelsea, and were separated at the age of eighteen months. "A" remained at Chelsea, and "B" was taken to a small town in Ontario. At the age of eighteen years "A"

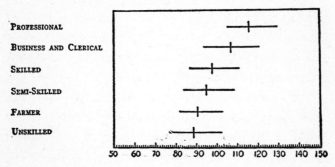

Fig. 12.—*Intelligence quotients of children of different social classes in New York State rural schools. The vertical line represents the median quotient. The horizontal line shows the range within which the middle 50 per cent of intelligence quotients lie.*

(*From Lorimer and Osborn 1934.*)

came out to Canada; a year later they were examined by Professor Newman. He found a remarkable difference between them. According to the Stanford Binet test the Canadian girl had a mental age of one year and eleven months more than her sister. By the International Group test the Canadian sister scored 62 per cent more than the Chelsea sister, and on all tests the

Canadian was the better. We might ask (but we shall not obtain an answer) whether the order would have been the same had the tests been designed in England. It is quite certain at any rate that such examples as this show that the results of intelligence tests do not depend entirely upon nature, but to a considerable extent on nurture. In eight other cases examined by Newman two differed more than this pair of girls, the difference in mental age being over two years; one pair showed about the same difference; the majority of pairs were very like one another. On the whole they were more alike than pairs of children chosen at random, and about as like as ordinary brothers and sisters brought up in the same environment. Clearly, environmental differences may have a large effect on the mental performance as measured by the intelligence quotient. Although we may criticise these tests, we have no better measure of intellectual development.

We now ask ourselves: What are the factors which are most influential? Probably the most satisfactory answer comes from the studies by Freeman, Holzinger, and Mitchell of adopted children; they revealed a considerable effect from a good home environment, enough to cause a marked resemblance of the adopted children to their foster brothers and sisters in the same home. They concluded that the school environment was less important than the home environment. It must, however, be remembered that the schools in the United States are more uniform than those in England. Other studies seem to confirm that school is less important than home; and that illness, accident and malnutrition [1] are relatively unimportant apart from a few well-defined illnesses which affect the brain, such as encephalitis lethargica. The matter is discussed in considerable detail in Schwesinger's *Heredity and Environment*, and in Gray's *The Nations' Intelligence*, a book which is none the less excellent because its author makes no attempt to conceal his political opinions.

Given that the environment has a certain effect we can ask ourselves whether all the differences between social classes which are measured by the intelligence quotient of the children are due to environment. The answer is that they are not. Lawrence's work on illegitimate children in institutions showed some correlation between the intelligence of the children and the social position of the parents. The position was similar (except in one research) with regard to orphan children brought up in orphanages from an early age. The differences between means of classes was about one-third of that observed when the children were in their home environment. It is, however, noticeable

---

[1] This is, of course, disputed.

that the differences in the intellect of school children are mainly due to relatively small classes at the top and bottom, namely the professional classes and unskilled and casual labourers.

Apart from the professional classes—doctors, clergymen, teachers, and so on—the well-to-do do not have children appreciably more intelligent, as judged by these tests, than the remainder of the population. Gray and Moshinsky found no difference between the children of officers and other ranks in the army. It may be that the differences which are found are largely due to the fact that at the present moment a certain type of intellectual achievement is fairly well rewarded in this country, and that the people who devise intelligence tests happen to be of this particular type of intellectual eminence. However, we may, I think, if the existing differences in fertility of social classes continue, expect a slow decline perhaps of 1 or 2 per cent per generation in the mean intelligence quotient of the country. That is, on the whole, deplorable.

Nevertheless, if one desired to do so, one could state another side of the question. One would suggest that granted this difference in "intelligence," it may be true that there are certain desirable qualities which are commoner among the more fertile part of the community. One might consider that the present state of the world was largely due to undue aggressiveness, that it was on the whole the more aggressive people who rose in the social scale; and that, in so far as aggressiveness was innate or dependent on innate factors, it might be desirable to humanity that those innate factors should be weeded out even at the cost of a certain sacrifice of innate factors making for intelligence. I do not think such a statement has ever seriously been made, or that the data in existence would warrant it being made; but I wish to point out that there is a possibility of a defence of the meek as being perhaps, in their way, not bad citizens.

There are, however, certain facts which lead me to doubt the truth of the deduction that, given the existing differential fertility, the mean (or rather median) intelligence of our population must be declining. If the eugenist argument is correct, a society in which men who rose by their abilities married a number of wives, while many of the poor went unmarried, would inevitably enjoy a slow but steady increase in intelligence.

Now this experiment has been tried, and tried with adequate controls. For more than a thousand years the Mohammedans in Western Asia have practised polygamy, whilst the Christians and Jews have not. Of course only the richer Mohammedans could afford a harem. We should therefore expect that the Mohammedans would on the whole be superior to the Jews and Christians in intellectual qualities, or at any rate in those qualities which make for the acquisition of wealth. In particular, a Turk should generally beat an Armenian or a Jew in a business deal. This is notoriously not the case. And because it is not the case it is to be presumed that there is some fallacy in the arguments as to the trend of our national intelligence which are drawn from the study of the differential birth-rate.

I am not going to suggest the nature of the fallacy. The whole question is enormously complicated, far more so than appears possible until one has studied genetics. If animal genetics affords any analogy, future work is likely to reveal entirely unsuspected facts concerning the determination of human intellectual capacity. The whole matter will only be cleared up by very careful combined work by geneticists and psychologists, work which in its early stages will probably appear to be quite unnecessarily abstract and academic.

Many remedies have been suggested for the differential birth-rate. A number of eugenists take the view that we should not tax the rich to educate poor children, many of whom could not benefit from education. Gray, on the other hand, in *The Nations' Intelligence*, points out that though high intelligence quotients are commoner among the children of the rich than of the poor, the majority of children of high intelligence come from elementary schools. Unless we can be certain that the suggested steps would greatly increase the fertility of the rich, they would therefore involve a considerable sacrifice of potential ability.

An even more extreme point of view has been put forward by a number of eugenists, among them being Professor Macbride, who in *Nature* (1936) writes: "There is only one remedy for the over-production of children that we can see, and it is very unpopular, so that it will probably be some time before the necessity for it forces itself on the public mind. This is compulsory sterilization as a punishment for parents who have to resort to public assistance in order to support their children."

Another remedy is available without legislation. Major Eric Suchsland, of the German Air Force, points out [1] the eugenic effect of air warfare for the following reasons: (1) bombing will be concentrated on regions where population is densest, poorest, and hence most undesirable eugenically; (2) during air-raids thieves will tend to come out, as well as anti-Fascists and other genetically undesirable elements intending to foment disorder; (3) genetically inferior people will manifest previously latent nervous and mental diseases, and thus become less likely to repro-

[1] *Archiv fur Rassen-und Gesellschaftsbiologie*, 1936.

duce their kind. I cannot find this paper at all humorous because I have seen its principles applied by German airmen to the improvement of the Spanish race.[1]

R. A. Fisher, in *The Genetical Theory of Natural Selection*, puts forward the view that in our existing economic system, apart from luck, there are two ways of rising in the economic scale; one is by ability, and the other by infertility. It is clear that of two equally able men—one with a single child, and the other with eight children—the one with a single child will be more likely to rise in the social scale. He may, for example, be able to save money and buy a small shop, and later to become relatively rich. Fisher points out that this was by no means so in the past, when a large family might be an asset to an artisan working at his trade in his own house. There is no question that people tend to marry into their own economic class. In the richer classes, according to Fisher, you have a

---

[1] When I wrote the above I had some doubt as to the propriety of linking Professor Macbride's name with that of Major Suchsland. Professor Macbride has since removed my doubt by using the columns of *The Times* for propaganda of German origin directed against the Spanish Republic. The statements which he reproduced from his German informant did not accord well with the facts which I personally observed in Spain. Such propaganda is of value to the German Government by justifying to the people of Britain the wholesale murder of Spanish civilians by German airmen. In less civilized countries eugenical arguments are employed for the same purpose. The views expressed by different writers on human biology and on questions of right and wrong are so closely related that it is necessary to examine the factual bases of their biological views (including, of course, my own) with very great care.

concentration at the same time of genes making for high ability and genes making for infertility. This result of our social system was first pointed out by Sir Francis Galton, who investigated the fertility of heiresses, an heiress being usually someone with no brothers and few sisters. They were found to have fewer children than the average. He held that the extinction of noble families was largely due to the practice of marrying heiresses. Thus the genes which had been in part responsible for the rise of the founders of these noble stocks were eliminated. Wagner-Manslau reached similar conclusions from a very detailed study of the German nobility.

Fisher hopes that this tendency may be combated by a scheme of family allowances. He believes that if wages or salaries were increased by 12 per cent for every child born the actual economic level of a family would not be lowered as a result of increasing the number of children in it. It is not certain that 12 per cent is sufficient; other authorities have doubted it. I have seen the figure put as high as 25 per cent. Whether or not such a scheme is possible for the country as a whole under our existing economic system or anything like it I do not know. It has been applied in France and Belgium in a number of industries. How far it has succeeded in checking the fall of the birth-rate is doubtful; it has certainly not checked it completely. On the

other hand, it has not fallen so rapidly in these countries as in England. My own view is that some form of
family allowances would be a measure of social justice,
but would be extraordinarily difficult to carry out. It
would be most difficult to arrange a system of family
allowances for the professional groups such as lawyers
and doctors, who are a valuable class of men from the
eugenic point of view.

Although I am in considerable sympathy with
Fisher's views, I cannot regard them as resting on very
complete evidence. It has been shown that within a
given social group (e.g., the English or German nobility) fertility is inherited in the sense that children of
fertile parents are themselves more fertile than the
average. But such resemblances may be due to example
and tradition rather than to a biological cause. Among
the data which would be needed before we could con-
clude that fertility is biologically inherited to a large
extent, that is, determined mainly by nature rather than
nurture, would be the following.

Monozygotic twins could be compared as regards
their fertility with ordinary sibs or with dizygotic
twins. It might well be found that the monozygotic
twins resembled one another in this respect to a much
greater degree than ordinary twins. This is certainly
true as regards stature, health, and intelligence. Professor Fisher demands an economic revolution. I am by

no means opposed to the idea of an economic revolution. But I think it rash to base such a revolution on the result of half a dozen investigations.

Brewer and Muller advocate eutelegenesis, that is to say, the procreation of particularly gifted men by artificial insemination. They believe that many married couples would willingly agree to the wife producing at least one child by a highly gifted father. Muller has ably urged this view in *Out of the Night*. Once again I am inclined to regard such a proposal as possibly premature in view of our very slight knowledge of the genetical basis of those characters which are found in the "great men" whom we regard as admirable.

Nevertheless I do not regard it with horror or disgust. It seems to me a far more desirable proposal than the compulsory sterilization of large classes. Compulsory sterilization would be a curtailment of human liberty, eutelegenesis an extension of it. It is perfectly conceivable that, when our knowledge of human genetics is sufficiently extended, some measure of this type may commend general approval. But, at least in Britain, I think that at the present moment we still possess a considerable store of mute Miltons and guiltless Cromwells. And as a means of producing more great men, equality of opportunity is more likely to be of immediate value than eutelegenesis. But for all that, I take the view that a couple who wish to practise this peculiar form of adoption should be at liberty to do so. However, like voluntary sterilization, eutelegenesis is clearly open to grave abuse. Thus a rich man who wished to have a number of children might offer money to women to have children by him, or offer to leave legacies to such children. Brewer's detailed proposals contain fairly adequate safeguards against such abuses, but this and other possibilities cannot be ignored.

My own views on the differential birth-rate are extremely tentative, but they are somewhat as follows: If the rich are infertile because they are rich, they might become less so if they were made less rich. A uniform and free school system, although it might be bad from the educational point of view, would probably be good from a eugenic point of view, as parents would not restrict their families to give their children a good education. I am inclined to believe inheritance of wealth eugenically undesirable, because it tends to make the well-to-do limit their families.

Curiously enough, the inheritance of property has been defended on genetical grounds. It has been claimed that a man who makes a fortune is abler than the average, and that it is desirable that property should be in the hands of able men and women, who will utilize it to the best advantage. Further, the children of able men are abler than the average, and should

134

therefore inherit the property. Let us, for the moment, accept the validity of this argument, and examine its consequences quantitatively. The correlation between parents and offspring as regards intellectual attainments is about half. This is partly due to direct heredity, partly to indirect heredity determined by the fact that the spouses of intelligent people are, on the whole, rather more intelligent than the average, and partly to environmental influences.

A correlation of half means that given the parent's intelligence grade, the variance (mean square of deviations from the mean) of the offspring is reduced to three-quarters of its value in the general population. In other words, one-quarter of the causes making for variation have been removed. On this basis it would seem fair that one-quarter of the property should be handed down, i.e., that death duties should average 75 per cent. I do not put forward this argument very seriously. But it seems to me about as good as the opposing argument. Both are examples of the impossibility of passing directly from biology to economics. I hold that the biologist can, at best, only suggest the probable biological effects of a given economic measure, and that even here he will very often be mistaken.

But I fully realize that these views are not held by the majority of my colleagues in this country who have studied the question. In my opinion it is impossible to arrive at any views on the problem which can at all reasonably be called unbiassed. One is inevitably biassed by political and economic opinions which are determined by facts other than one's biological knowledge. Five or six hundred years hence people may be able to judge what would have been the correct policy for their forefathers from the eugenic point of view, just as we can speak wisely on the weaknesses of feudalism. I do not think that we are able ourselves to make such a judgement because, taking a part in the economic struggles of our own time, we cannot be as impartial as if we were considering the eugenics of domestic animals. We are part of history ourselves and we cannot avoid the consequence of being unable to think impartially. I hope, however, that I have shown that this whole problem links up with a great number of wider political questions. Ever since the time of Plato the different innate endowments of different classes have been a matter of political speculation. I do not believe that any of these eugenical schemes are likely to be of much importance because I take the view that the economic changes which we may expect in the near future will be determined by causes much more powerful than the arguments that any biologist may bring forward; and it may be desirable that biologists should confine themselves to questions such as the inheritance of well-marked characteristics concerning which it is possible to arrive at some measure of agreement. If they do not they may prejudice large sections of society against whole fields of biological research. Just because my own views differ from those of many of my colleagues, I feel myself fully justified in giving them publicity, if only to make it clear that a consideration of human biology does not, in my opinion, justify the perpetuation of class distinctions. If this view were shared by all, or almost all, human biologists, I should be much more inclined to confine myself to the academic aspects of science, and to leave to others the discussion of its political implications.

CHAPTER V

The Nature of Racial Differences

THERE is a very general belief in the existence of racial differences in psychology. For example, it is commonly said that as compared with whites, Negroes are more musical but less intelligent. Similarly it is stated that Japanese are more artistic but less honest compared with Englishmen (I am only repeating and not endorsing statements of the kind frequently made) and that compared with the English the Irish are more cheerful but also more pugnacious. It is further said that differences of this kind are inborn and that no amount of education will altogether remove them.

Such beliefs have been generally held for a long time. We are so accustomed to hearing of the superiority of Europeans that it is perhaps worth quoting [1] from the Moorish writer Said of Toledo, who wrote at the time when Toledo was in Moorish hands. Describing the people living north of the Pyrenees—our own ancestors—he said: "They are of cold temperament and never reach maturity. They are of great

---

[1] After Hogben, *Genetic Principles in Medicine and Social Science.*

138

stature and of a white colour. But they lack all sharpness of wit and penetration of intellect." We must remember that seven hundred years ago such a point of view had at least an empirical justification, for at that time trigonometry was being studied in Toledo, while in Europe a man was regarded as learned if he had got as far as the fifth proposition of the first Book of Euclid.

The main tendency to deny this thesis has come, I think, from the supporters of movements which claim a universal applicability to the whole human race, such as Christianity, Islam and Marxism. A good Christian should believe that a Christian Negro is better in God's sight than a white infidel, and presumably he will act accordingly. Now there is no question in my mind of the existence of racial differences in psychology, but I do not know to what they are due. It is often claimed, for example, that the low intellectual status of many non-European races is due to the fact that they have been conquered, that they have been denied access to European culture, and so on. There is an element of truth in this; nevertheless the differences between the reactions of different races to European culture are extremely startling. Let us consider the state of affairs in two neighbouring countries, Australia and New Zealand, which have been colonized by very similar whites but which contain very different aboriginal races.

The Maoris in New Zealand belong to the Polynesian race. There is no question that they have managed to co-operate to a very considerable extent with the European immigrants. Perhaps the most remarkable example of this is to be found in the career of Mr. Pomare. Mr. Pomare, who, as his name implies, was a Maori, became a Minister in the New Zealand Cabinet, and for a certain time was actually acting Prime Minister of New Zealand. The whites found him no worse than other Prime Ministers.

We may contrast this with the condition in Australia where the aboriginals are of a black race quite different from the Negroes of Africa. We know that in Australia there has been a complete failure to assimilate the blacks into the European culture, that none of them has ever risen to any rank of the least importance in Australian society, and that as a result the racial future of the two countries is likely to be very different. There is extensive crossing between whites and Maoris in New Zealand, and it is likely that in the course of a few centuries New Zealand will be inhabited by a race of mixed origin. On the other hand, unions between whites and blacks in Australia are illegal, and it is quite possible that the black race will ultimately become extinct. It is unlikely that it will contribute to any great extent to the final population of Australia. Yet we must remember that the Maoris put up a fierce resistance to

British conquest, that they were beaten in a series of wars, and that many of their practices, such as cannibalism, were of a kind definitely abhorrent to their conquerors. We cannot, I think, deny a very considerable difference in the behaviour of the Maoris and the Australian black-fellow, and we can ask whether it is due to nature or nurture? That is an exceedingly difficult question to answer. But I find it very hard to rule out nature. Granted, however, that there such racial differences exist, we can ask the very important question whether there are differences of the same kind between different peoples living in Europe. For example, should we be in any way justified in arguing from the difference between the Australian black-fellows and the Maoris to a difference, not so great perhaps, but of the same kind, between Germans and French, Germans and Jews, Nordics and Alpines?

Before we can begin to answer that question we must try to examine the meaning of the word race. This is a word which we use very freely and to which we attach, perhaps, more meaning than is wholly justifiable. It will be agreed, I think, that the differences between races, or at least the differences by which we define races, are innate differences, differences of nature and not of nurture. The Negro in England does not assume the English race if he becomes an English-speaking British subject and a member of the Anglican

Church. On the other hand, if we tried to separate the human species on the basis of their innate physical characters we should not always get a division into races. For example, if we consider the various black-skinned, kinky-haired, thick-lipped people they would all be Negroes. On the other hand, if we separated out all the freckled, red-haired people we should find that they were a section of the Europeans, commoner in some parts of Europe than in others, but nowhere constituting the whole or even the majority of the population. In the same way there is not an albino race, an idiot race or a colour-blind race. Clearly, however, we cannot rush to the other extreme and define a race solely on the basis of geographical location. We might provisionally define a race as a group which shares in common a certain set of innate physical characters and a geographical origin within a certain area. I do not pretend that no criticisms of that definition could be made. It is important to realize, however, that if we tried to define race on the basis of either the physical characters alone or geographical origin alone we should be led into errors.

Other authors go a good deal further. For example, the German anthropologist, Günther, says: "A race is a human group all of whose members are alike in physical and mental characters." That is unsatisfactory for a number of reasons. It would tend to lead us to separate out, let us say, red-haired people as constituting a separate race, and if we thought it were true as regards some particular group we should tend greatly to underestimate the amount of variability which exists within every race.

It would be theoretically possible by inbreeding to produce a race all of whose members were remarkably alike in physical characters and closely resembled one another in their behaviour. But it would only be possible as a result of very prolonged inbreeding such as has actually taken place with many of the races of domestic animals. We should ultimately expect to reach a population all of whose members resembled one another as much as do a pair of monozygotic twins. Such twins resemble one another to a remarkable degree not only in physique but in character. For example, one such pair were both professional burglars, although they never on any occasion co-operated in a burglary as far as was discovered. The degree of resemblance in character which goes with the great physical resemblance between monozygotic twins has been conclusively shown by a series of studies of criminal twins in Germany.[1] If one of a pair of monozygotic twins is a criminal the other will more often than not be a criminal of the same kind, whereas in the case of ordinary

[1] The largest of such studies is that of Lange, translated under the title *Crime as Destiny* (Allen & Unwin, 1931).

twin pairs the large majority of brothers of criminals are not themselves criminals. If members of a given race resembled one another to that degree it is clear that from the study of a small sample we could predict pretty well the behaviour of the remainder of the race. It would be interesting to speculate what kind of political constitution would be suitable for a group of people all of whom possessed the same innate characteristics.

It is clear that such a definition as Günther's is considerably exaggerated. No race is homogeneous; and none is anything like as nearly homogeneous as highly inbred and selected races of domestic animals such as greyhounds, Jersey cows, or South Down sheep. We cannot legitimately argue from the experience of our domestic animals to the human race. We can say that in the case of the greyhound a particular physique is associated with the habit of hunting by sight rather than by smell, and so on. It may conceivably be that similar correlations will be found in the case of the human races, but we cannot use our knowledge of the domestic animals as the basis of any *a priori* arguments.

Let us try to make our ideas on race a little clearer by taking some concrete examples. Let us compare two random samples of a thousand Englishmen and a thousand Negroes from the Gold Coast. As regards certain characters there will be no overlapping between them whatever. The darkest Englishman will be a great deal paler than the lightest Negro. In all probability the curliest-haired Englishman will have straighter hair than the straightest-haired among the Negroes. Probably the same will hold as regards the thickness of the lips and several other physical characters. On the other hand, with regard to a great many physical characters, notably stature and skull shape, there will be a great deal of overlapping. It may be that the Englishmen will be on the average a little taller, but a great many of the English will be shorter than the average of the Negroes. Further we can be quite sure that the differences between them so far considered are mainly if not wholly due to nature and not to nurture. Even after living for several generations in a tropical country the colour of the skin of English people, though perhaps a little darker, will not begin to approximate to the colour of the palest Negro.

If we had only a few samples of the human species we could divide it with considerable confidence into a number of races with no overlap between them. If a naturalist from some other planet were to descend to this one and shoot a few specimens of men in various continents, when he returned and classified them in a museum he would very likely divide humanity into several different species. On the other hand, if he travelled widely over our planet he would find a distressing number of intermediates. If he travelled from England to the Gold Coast he would find a number of populations intermediate as regards ranges of colour, each of them overlapping with the last. As he travelled through Southern Europe and Syria, down the Nile Valley, and through the Sudan, he might be tempted to explain these intermediate populations as due to hybridization between originally distinct races. That would be, however, a hypothesis which would require further examination. However this may be we are wholly justified in regarding the English and the Negroes as members of distinct races. From examination of a single individual we can say with certainty to which race he or she belongs.

Now consider the case in Western Europe. There we find a number of different populations. We find that as we go southward the people become on the whole darker. Nevertheless, we do find a considerable degree of overlapping. Supposing we had a random sample of a thousand Swedes and another random sample of a thousand Sicilians. We should have no difficulty in saying which consisted of Swedes and which of Sicilians, for the Swedes would be mainly blondes and the Sicilians mainly dark. Nevertheless we could not with certainty assign any particular individual to one or other group. The darkest Swede would be considerably darker than the lightest Sicilian, and if we

classified our individuals on the basis of their hair and
eye-colour we should make a number of mistakes.
There would be a certain amount of overlapping as regards innate characters, and that would be so throughout Western Europe.

Nevertheless, attempts have been made to describe
a number of races within Western Europe. The Nordics are characterized by fair hair, blue eyes, large
stature and a long head; the Alpines are characterized
by medium hair, brown or blue eyes, and very round
heads; and the Mediterraneans by dark hair and eyes,
and long heads, but a smaller stature than the Nordics.
Now these types are of considerable value for classification. If we go to Sweden we shall find that a considerable proportion of the population agree pretty
well with the Nordic type. If we go to Switzerland we
shall find that most of the people show several of the
Alpine characteristics. If we go to Sicily we shall find
the majority of the people to be of the Mediterranean
type, and provided that we speak of a Nordic type or
an Alpine type we are perfectly justified. Nevertheless,
if we speak of a Nordic race we must remember that we
are using the word race in an entirely different way to
what we use it for when we are speaking of the Negro
race. As the word race is one to which a good deal of
emotion is attached, it is of considerable importance to
use it so far as possible in our discourse with only one

meaning, instead of attaching two or more different meanings to it on different occasions.

It must at once be admitted that the Swedes and Sicilians differ as regards their innate characters, but it must be remembered that the difference is a statistical difference allowing of overlapping between them, and not an absolute difference such as exists between whites and Negroes. The best criteria for the classification of Europeans on the basis of innate characters are the pigmentation and the ratio of breadth to length in the skull. Other skeletal measurements are by no means so good for diagnostic purposes.

When we examine a large population as regards any physical character we find that a certain measurement, generally close to the average, is most frequent, and that as we go above or below that the frequency falls off. Thus in a classification of two thousand women on the basis of their stature, the average stature was $62\frac{1}{2}$ inches and the frequency fell off rapidly on each side. The important measure called the standard deviation is the difference between the mean and the point at which the curvature of the frequency curve alters, in this case about $3\frac{1}{2}$ inches. Now supposing we take a number of European populations and measure their skeletons. In each case we can define the average. We say that the average or typical Englishwoman is $62\frac{1}{2}$ inches high, and so on, and we can make up an imaginary population of the typical members of a number of European races. Dr. Morant [1] made up such a population based on eighty-five different European groups and then asked himself how variable was that population of types. He found that as regards skeletal characters it showed standard deviations averaging about half those of any population from which it was derived. Thus the imaginary population of a typical Englishman, a typical Scotsman, and so on, is much more homogeneous than any real human population ever found in practice. In other words, the differences within a race are greater than the differences between races. The one skeletal characteristic on which you can divide up European populations is the breadth to length ratio or cranial index of the skull.

Fig. 13 represents the distribution of the cranial index in three European populations: Swedish, Russian and Piedmontese. The medians differ very markedly. However, there is a considerable degree of overlapping even between the Swedes and Piedmontese, who represent extremes. The Swedes have the narrowest heads and the Piedmontese the broadest. But even their distributions overlap to the extent of 20 per cent. If we drew a line in the most favourable place for making a distinction we should still assign 10 per cent of the Swedish skulls to Piedmont and of the Piedmontese

[1] Unpublished personal communication.

149

skulls to Sweden. Thus even in an extreme case of divergence in physical characters between European populations we find overlapping between different

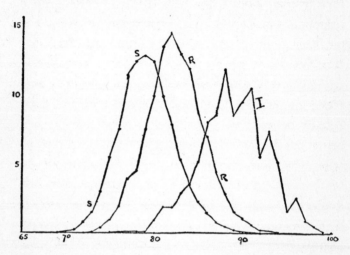

FIG. 13.—*Frequency distributions of cranial index (breadth as percentage of length), S in Sweden (after Lamborg and Linders), R in Russia (Smolensk district, after Tchepowkovsky), I in Piedmont (after Livi). Abscissa, cephalic index; ordinate, percentage frequency. The irregular form of the Piedmontese curve is due to the small number (1,238) examined. All measurements were made on living men.*

groups, and we must be very careful of speaking of races in Europe.

I will, however, examine for a moment the evidence for the existence of certain alleged races. There is, for example, the Jewish race. There is no question that as regards physical characters the Jews overlap the Europeans among whom they live. What is more, as regards their average skull shape they tend to resemble their neighbours in Europe to a considerable degree. That has been pointed out recently by the late Professor Karl Pearson in his last paper. He concluded that this resemblance suggests that the Jewish race is not by any means as pure as many both among Jews and anti-Semites believe, the resemblances to their neighbours being due to inter-breeding. The question therefore arises whether the undoubted peculiarities of the Jews, or to put it more accurately, the characteristics undoubtedly more frequent among Jews than among their non-Jewish neighbours, are due mainly to innate characteristics, or to the peculiar character of the Jewish religion and culture.

Even less convincing, in fact much less convincing, is the case for a German race. This expression is being used to a considerable extent nowadays for propaganda purposes. A case can be made out for the inclusion in the German State of a number of people living outside it who speak the German language and who feel a unity with the people living inside it. I do not intend to examine the arguments for and against this proposal. But when we examine the Germans physically we find that they are extremely diverse as regards

their innate physical characters. We find that in the
north, along the Baltic coast, there is a population
which is predominantly Nordic, while in the south, in
Bavaria, the population is predominantly Alpine, with
moderately dark hair, round heads, and so on. The diversities among the Germans as regards their innate
physical characters are very much greater than among
the English, but not much greater than among the
French, where one finds, for example, Nordics in Nor-
mandy, Alpines in Auvergne, and Mediterraneans in
Provence. Nevertheless, in spite of this diversity in
physical characteristics, no one would deny the very
real unity in behaviour and sentiment among French-
men from different parts of France; and there is no
need, when we find that the Germans are considerably
more diverse racially than the English, to suggest that
Germany should be broken up into a series of states
on a racial basis. Actually an attempt to divide up Europe on the basis of innate physical characteristics
would have the most remarkable effects. Both Germany and France would be divided into three or four
portions, the Baltic populations of Germany being
united with Scandinavia, the Frisians with Holland,
the Bavarians with Austria and Switzerland, and so on.
In fact, one may say at once that any attempt to divide
up Europe on a racial basis would make the Treaty of
Versailles appear as an absolute model of justice and

sanity. It is of some importance that we should realize that the real divisions in Europe are not primarily on a basis of race as definable in terms of innate physical characteristics, but on a basis of culture, language, and so on.

Other alleged races of which we hear a good deal nowadays are the Aryan and Semitic. These terms were first used in discourse in Europe to define families or groups of languages. The Aryan languages, which extend from the Atlantic to Northern India, have very readily recognizable characteristics in common, and are presumably derived from an original Aryan language. But among their speakers are included a great variety of people, not only the majority of Europeans, but dark-skinned people in Northern India, and Armenians who possess to an even greater degree than the Jews the hook-nose which is regarded as so characteristically Semitic. Fortunately or unfortunately the physical differences between human beings are much less marked than the differences between the languages which they speak. In the same way, although the Jews in many countries and the Arabs speak a Semitic language, nevertheless on the basis of a number of physical characters and particularly the blood groups, with which we shall have to deal later, the Jews resemble Europeans fairly closely and diverge to a great extent from the nomadic desert Arabs, who, however, presumably supplied a certain proportion of the Jewish ancestry and imposed a Semitic language on the whole people.

In view of these facts it is all the more remarkable that Semites should have another physiological character in common. Dr. Darré, the German Minister of Agriculture, writes: [1] "The Semites reject everything connected with the pig. The Nordic peoples, on the other hand, give the pig the highest possible honour. . . . The Semites and the pig are faunal and thus physiological opposites." Unfortunately the evidence for this profound physiological discovery has not yet been published.

We now come to the question of psychological differences between the races. Perhaps the first formulation of such differences was made by de Gobineau. His theory was that sensory acuity was highest in Negroes, intermediate in Mongolians, and lowest in Europeans. The Europeans, however, compensated for this by being somewhat more intelligent than the Mongols and considerably more so than the Negroes. Intelligence is not a very easy thing to measure, but the acuity of the senses is quite easily measured, and unfortunately the examination of a number of primitive peoples has shown that they do not possess this remarkable sensory acuity as judged by any tests which psychologists

[1] "Das Schwein als Kriterium der Nordischen Rasse."

154

have yet been able to devise. The most striking examination of this kind was perhaps made by Woodworth at the St. Louis Exposition in 1904, where there was a world congress of races, and representatives of many primitive peoples attended. The whites were on the average superior to everybody else, the one exception being a small group of pygmies from the Congo who were particularly adept at hearing very high notes. The superiority of the whites was probably very largely due to the considerable amount of eye and ear disease among the primitive peoples which had not received any adequate treatment.

As regards intellect and character, however, tests are very much more difficult. If I stated that race A is superior to race B as regards some particular characteristic, say musical ability, I might mean one of four different things. First of all I might mean that there was no overlapping between them whatever. I might mean that the least musical Negro was more musical than the most musical European, or that the cleverest Negro was stupider than the stupidest European. I think it will be generally conceded that no examples of that kind are to be found. No one has yet discovered any psychological trait in which there is no overlapping between races.

Secondly, I might mean that the inferior race had some upper limit to its achievement placed on it by

innate characteristics. It has been said that all the brains
of bushmen so far examined are of a decidedly primitive character morphologically, and that therefore it is
impossible for the bushmen to show more than a modest amount of intellectual achievement. It has also been
urged that the microscopical anatomy of the brains of
East African Negroes is different from that of Europeans and inferior, and that therefore not much can be
hoped for from educating such races. In the case of
the bushmen the difference is presumably innate. In
the case of the East Africans it may conceivably be
due to the very inadequate diet which they have been
shown to enjoy. Nevertheless, the correlation between
observable brain structure and performance is so slight
that we must be careful of building too much on arguments of that kind unless they can be supported by
substantial evidence based on the achievements of large
numbers of members of the two races.

Thirdly, I might mean that although there is overlapping, the average or median performance of one
race is superior to that of the other. If we are classify-
ing people as regards something which cannot be actually measured, such as intelligence, we cannot really
speak of the average intelligence. We can speak of the
median intelligence. Supposing I arrange a large number of people in a row and place them in order of their
intelligence as judged by a particular test. I can pick

out the one in the middle, and he acts as quite a satisfactory representative. We may claim then that although there is overlapping as regards intelligence between Europeans and Negroes, the median European is more intelligent than the median Negro.

The fourth possible meaning is that, as regards the character concerned, exceptionally gifted people are very much more frequent in one race than in another. For instance, when we say that the Germans are a musical people we are probably thinking of Bach, Beethoven, and Wagner, rather than of the average German. We must remember that as regards many kinds of human achievement, it is only the exceptional individuals who count. We know that the Greeks produced a number of great mathematicians, we have no evidence that the average Greek was better at his arithmetic than the average Roman, possibly quite the contrary.

Opinions differ to a very great extent as to the importance of exceptional individuals in the cultural and political development of a people. That is not a question which I propose to take up here. I would merely point out that it is quite conceivable that some primitive peoples who have made no great contribution to human culture may have been handicapped rather by homogeneity than by a low average. One might expect to find a great homogeneity under certain conditions, which would mean that exceptions either upwards or downwards from the average were comparatively rare.

Further, if we observe a difference between two races, it may be due to nature, to nurture, or to their interaction.

Let us now look at a few examples. Large numbers of intelligence tests have been made in the United States. It is found that the Chinese and Japanese do practically as well as the whites on the average, and seem to diverge about equally from the average in both directions. On the other hand Negroes, who, of course, in the United States are almost all hybrids, and Red Indians, do very considerably worse, the average intelligence quotient of both groups being even as low as 75 per cent of that of the whites.

We have next to ask whether these differences are due to nature or to nurture. Striking evidence on that question is provided by the intelligence tests through which the recruits for the American Army in 1917 were put. In the army alpha intelligence test in any given State, the median value of the whites was invariably higher than that of the Negroes—generally 15 or 20 points above it. In five Northern States the whites averaged 60 and the Negroes 40, while in eight Southern States the whites averaged 40 and the Negroes 20. That is to say, in the Northern States the Negroes did as well in the intelligence tests as the whites in the

Southern States, and the Ohio Negroes beat the Arkansas whites by 10 points.

From that result one of two things seems to follow necessarily. Either cultural differences play a greater part in determining the response to these intelligence tests than has been generally believed, or alternatively, selective influences within a race can alter its character to a very great extent. It has been suggested that the more intelligent whites and Negroes leave the South to go North. That may be true to some extent and it may be that the intelligence so selected was innate. If so, it shows that selection can have a very large and rapid effect on the intelligence of a race and that therefore racial intelligence is not fixed.

My own view is that probably there would be slight differences found in the results of intelligence tests if these people were brought under a precisely similar environment, but I should hesitate to say in which direction they would be found, except to suggest that as the intelligence tests have all been devised by whites, they would be likely to show a certain superiority of whites over Negroes.

Davenport and Steggerda tried to get over that difficulty by examining the whites and Negroes on a small island in the West Indies living under very similar cultural conditions. Each race was found to be superior in certain respects, though the authors think that on the whole the whites did better. In order to find a white community comparable with a Negro community they had to go rather low in the social scale so far as the whites were concerned, for while several of the whites had been to gaol for petty offences, several of the Negroes had even been to college. The fact that they found the Negroes rather more superior to the whites in some respects than they had expected may perhaps be attributed to that. What I want to point out is the impossibility of getting a fair comparison in a society where the members of races are differently treated. If you take the population as a whole the conquered or exploited race will undoubtedly be handicapped; if you take a section of the population where their economic conditions are much the same you will get a very poor sample of the conquering race. Therefore I can only close this question of the alleged superiority of whites over Negroes on a note of agnosticism. I can state the view that not merely has nothing been proved, but that it is going to be exceedingly difficult to prove anything within the next few generations.

When it comes to different European races there has been a tendency in America to regard the Nordics as very superior, and there is little question that on the whole immigrants from "Nordic" nations do better in the United States than those of other European racial groups. On the other hand, when tested in their own environments the differences are very slight. Predominantly Alpine Paris is practically equal to

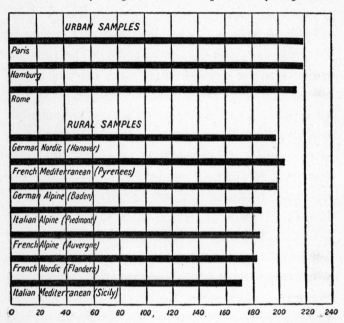

	URBAN SAMPLES											
Paris												
Hamburg												
Rome												
	RURAL SAMPLES											
German Nordic (Hanover)												
French Mediterranean (Pyrenees)												
German Alpine (Baden)												
Italian Alpine (Piedmont)												
French Alpine (Auvergne)												
French Nordic (Flanders)												
Italian Mediterranean (Sicily)												

0   20   40   60   80 100 120 140 160 180 200 220 240

FIG. 14.—*Average scores of groups of 100 boys in an intelligence test. The boys and their parents were born in the area in question, and in the rural areas only those boys were tested who showed the physical characters regarded as typical of the race in question. From Lorimer and Osborn, after Klineberg.*

predominantly Nordic Hamburg and superior to Mediterranean Rome (Fig. 14). But another group of Mediterraneans from the French Pyrenees did better than the other rural groups, and a Nordic group in French Flanders was the worst but one. One cannot say on these data that there is any evidence of the superiority of one "race" over the other.

When it comes to the percentage of exceptional individuals we get rather surprising results. Fig. 15 gives the results of Terman as to the frequency of gifted children in a Californian population in proportion to the frequency with which they were to be expected if they were equally frequent in all immigrant groups. There is an enormous excess of children whose parents came from Russia. That, let me remark at once, is not a tribute to Communism, but to Judaism, the majority of the emigrants from Russia having been Jews. It can be argued that the surprising number of intellectually gifted children is due to the innate superiority of the Jews. It can also be suggested that it is due to the fact that Jews mature earlier intellectually, or it can be put down to the characteristic cultural environment of the Jews. Similarly the low showing made by Irish, Italians, and Mexicans may be put down to the fact that most of the emigrants from those countries were unskilled labourers who had low innate endowments, or who brought up their children under such conditions that they were not likely to do well in intelligence tests.

A certain amount has been done on the scientific

testing of temperament. I should be inclined to expect larger temperamental differences than intellectual differences. There is a great deal of overlapping, but there are certain differences in average. For example, in re-

FIG. 15.—*Frequency of gifted children in different foreign-born groups in California. From Lorimer and Osborn, after Terman.*

gard to the repeated performance of a task intended to be fatiguing, the Negroes tested by Garth [1] showed less persistence than the whites. Indians from the American plains started poorly but showed a good deal

[1] *Racial Psychology* (McGraw-Hill).

of persistence. It is, however, necessary to emphasize
the extreme degree of overlapping. One cannot iden-
tify an Indian or a Negro by the type of performance
which he gives. Still less can one recommend, on the
basis of these studies, that Negroes as such should be
excluded from a particular group of careers. At most
one can say that they are slightly less likely to succeed
than whites or Indians. And there is no evidence as to
whether their difference from the other races is to be
attributed to nature or to nurture.

CHAPTER VI

The Race Problem Continued. Conclusion

I WANT to examine for a moment the doctrine of the fixity of racial characters which plays so considerable a part in teaching in Germany to-day, where it is even suggested that the Nordic race has somewhat deteriorated since the Stone Age, but may hope to get back to its original heights. Probably the only European country in which we can study not indeed a pure race but a race practically "uncontaminated" for the last two thousand years is Scandinavia. There have been many invasions of other countries from Scandinavia; there has been very little immigration into it except of a few nomadic Lapps into Northern Sweden. If we examine the characteristic of the Scandinavians one thousand years ago we find that they were a race of skilled warriors and efficient pirates. I believe that in the palmy age of the Vikings there is no record of a Viking defeat by any odds of less than ten to one. If we went there nowadays we should find that politically they were pacifists, that they went in for a rather mild form of Socialism based as much on co-operation as on State ownership; that they did not even

165

practise the more militant forms of Socialism such as Communism. If we had no experience of the courage and efficiency of Scandinavian seamen we might be led to suppose that in spite of preserving their Nordic blood, they had degenerated and become completely cowardly as they have become to a very large extent unwarlike. That example shows the very great difficulty of judging the fundamental characters of a race from its institutions. I am inclined to believe that the Scandinavians are innately at least as courageous as the rest of us—it may take a certain amount of courage to be a pacifist at times.

The only probable way in which the racial stock could have changed would have been by some form of selection. It has been argued that the more enterprising Scandinavians left the country, and that the alleged degeneracy is due to that fact. That brings one up against a rather interesting contradiction between two ideas which exist together in a state of uneasy juxtaposition in the minds of some of the people who believe in the extreme importance of innate characters. One is the idea of the fixity of racial characters, a doctrine which is of course incompatible with Darwinism. The other is the idea that a race may rapidly degenerate as a result of the survival of the unfit, and be rapidly purified as the result of some eugenic measures. Both of these notions can hardly be true.

If this example is not conclusive, an instructive experiment may be tried. Take any of the more blood-stained pages of the Anglo-Saxon Chronicle,[1] in which aldermen's heads fall like so many ninepins. Read it aloud to a friend, substituting Kavirondo for Kent, Mbonga for Eadfrith, and so on. Then ask him whether he thinks it even remotely possible that in five hundred years the descendants of these bloodthirsty savages will be building some of the world's most delicate architecture and laying the foundations of a remarkably just and stable form of government. If he says yes, you will probably find that he is either a firm supporter of Christian missions or a Marxist.

I now come to the theory of original pure races. For example, it is sometimes said that the existing Germans are composed of 60 per cent Nordic, 20 per cent Alpine, 15 per cent East Baltic, and 5 per cent Dinaric, or some other numbers, and that they are to be regarded as a mixture of original pure races in those proportions. No doubt the existing statistical situation could be expressed to some extent in terms of a mixture of hypothetical units which were once pure, just as one can express the composition of seawater as so much distilled water, so much sodium chloride, so much sodium sulphate, and so on; but any chemist will realize that the particular set of salts which make up this hypothetical

---

[1] Translation in Dent's Everyman's Series.

seawater is quite arbitrary, that all the sulphate might have been present as sodium sulphate or some as potassium sulphate, the lithium as chloride or bromide, and so on. In the same way with regard to original hypothetical races, one can represent the composition in a good many different ways. When we come to actual evidence as to whether there ever were any pure races, we have to remember that our only scientific data are based on the study of skeletons. We have a few remarks by historians to the effect that the Germans had light hair. It has been assumed on that basis that all Germans two thousand years ago had light hair. That seems to me to be a rather big assumption to make on the basis of single sentences by authors who had probably not studied very large numbers of Germans in their native forests. But when it comes to skeletons we find with a number of ancient populations that although their average values for physical characters may sometimes be fairly extreme, as in the case of the long-headed Neolithic people in England, yet the variation is just as great in the primitive groups as in modern populations. That is so of the "Nordic" people of the *Reihengräber* in Germany, the pre-dynastic Egyptians, and many others. There seems to be no real evidence for the suggestion that the variation in stature or skull shape in the modern English population is due to racial mixture, but there is a certain amount of evidence that the variation in pigmentation is so. The pure Nordic race of the past is at worst a myth, at best a deduction from inadequate evidence.

When we come to consider more modern populations we have to rely to a remarkable extent on the work of one or two people and particularly on that of Dr. Morant, of University College, London. He has spent much of his lifetime in the exact measurement of large numbers of skulls and in somewhat intricate statistical treatment of the results, so intricate indeed as to be inaccessible to many readers. One of his fundamental ideas is what he calls the racially homogeneous population, such a population as would be provided by the agricultural labourers of, let us say, an average English Midland village. Within such a population the variation is as small as you can ever find, and very considerably smaller than in the total population of any large country or of a continent such as Europe.

What is more interesting is this. Within a racially homogeneous population certain different physical characters are not to any appreciable extent correlated. I will explain what I mean by correlation. As we go through Germany from the north to the south we start with a predominantly fair-haired and long-headed people, and in Bavaria end up with a predominantly brown-haired and round-headed people; and if we took one thousand Germans from various different areas we should find a correlation between fair hair and long heads. Among the people with very fair hair we should find on the average greater height and a longer skull than among the dark-haired people, and we should find the same thing in England though to a less extent. We should find an association between fair hair and long heads on the East coast, where there were many Nordic invaders; but in the population of a dozen Oxfordshire villages we should find no correlation of that sort. That is entirely intelligible to the geneticist. It means that head shape and hair colour are determined by different genes. For example, if we had a number of rabbit populations all of which were the result of crosses between coloured, short-haired rabbits and white long-haired "Angora" rabbits, we should find that in some of these populations there were very few white rabbits, and in these same populations we should find very few long-haired ones. But in any given population, say one produced by the crossing of equal numbers of dark short-haired rabbits and white long-haired rabbits and the breeding together of their descendants, we should not find any correlation. We should not find long hair any more likely to be associated with whiteness than with colour, because they are due to different genes that are very readily dissociated.

This is of some importance because it means that the

physical characters of races are due to a number of different genes which can be separated, and it follows that a racial mixture once made is to a very large extent irreversible, and cannot be undone on the basis of selection for some particular characters. If there was once a Nordic race which carried genes for great courage and initiative it is extremely unlikely that those genes were the same as those for the blue-eye colour and long-head shape which accompanied them. It is surely probable that the initiative and other admirable qualities, if they were genetically determined, were determined by different genes. One is particularly led to such a belief by a rather curious fact in animal breeding. If we examine many breeds, for example, breeds of cattle, we shall find that each breed has a standard colour; but if we take the one animal which has had the most intense selection of any, this is no longer true. In the case of the thoroughbred English race-horse, which has been very intensely selected for running power, you will find a fair number of colours. It is at least probable that in so far as there are large innate psychological differences between human races, they are not mainly due to the same genes which produce the colours and other characteristic differences. This is of some importance, as it suggests that colour prejudice or the colour bar is an inadequate substitute for race prejudice or the race bar.

A certain amount of light is thrown on the nature of the differences between the major races by the results of crossing them. I shall for the moment only consider those characters which can be definitely measured or observed, and which depend very little on environment, that is to say, a certain group of physical characters.

The first generation in such a cross is generally intermediate, but may resemble one parent rather than the other. Here we must be wary of subjective impressions. We can readily detect a relatively slight amount of Negro ancestry, "a touch of the tarbrush," as it is called. A pure Negro could probably detect "a spot of whitewash" which we should not notice, just as at first we do not notice differences between Negroes for which a Negro is on the lookout.

In crosses between Europeans and West Africans, such as those which furnished most of the coloured population of the United States, the skin colour is roughly intermediate, but the hair form is predominantly that of the Negro. The same is true of crosses between whites and the Bantu-speaking negroids of South Africa. But with other black races things are very different.[1] The first generation of the cross between Europeans and Melanesians commonly have

[1] In the lecture course slides were shown to illustrate these points.

172

wavy—not even curly— hair of the European type, and the skin may be pale-yellow rather than brown.

In later generations there is no general rule. The union of a white and a mulatto (first cross of European and West African) rarely if ever gives anything like pure whites. The racial characters blend. That is to say, they are probably due to a large number of genes on different chromosomes. Thus supposing there were genes for colour on ten different chromosomes of the West African, we should expect one in a thousand from the mulatto-white cross to be white (though probably with negroid hair or lips), and only one white in a million from the union of two mulattoes.

But with the Bantu negroids the work of Lotsy and Goddijn has shown a very different situation. Marriages between Europeans and half-castes of the first generation often give children with white skins, blue eyes, and straight yellow hair. This presumably means that the South Africans differ from Europeans as regards a small number only of genes making for colour and hair form.

The same is true of the Chinese. Among the children of two parents each derived from a Cantonese mother and a European father, appeared one boy with grey eyes without the epicanthus which gives the characteristic Mongolian slant, brown hair with a gold glint and a pink-and-white complexion. In fact, he could pass as an English boy, and is decidedly less "coloured" than most Italians or Spaniards. But at least one of his brothers was, at any rate to European eyes, definitely of the Chinese type. This again means that the Chinese, at least as regards the more obvious physical characters, differ from Europeans in respect of a few genes only. In fact it is not infrequent for a hybrid of the first generation to produce a gamete with a set of genes characteristic not merely of Europeans, but of northern Europeans. And when two such gametes unite we get a child of northern European type.

The social consequences of these facts are interesting. So careful an observer as Mrs. Millin notes that in South Africa the colour bar is not effectively a race bar. Men and women with negroid or Malay ancestry are often of European appearance, and are accepted as whites. Indeed several men who have achieved great distinction are commonly believed to have a certain amount of such ancestry, and perhaps they are none the worse for it. In the United States, on the other hand, the colour bar is much more effective in keeping men with Negro ancestry from high positions. We may note in passing that in the United States the very strong feeling in favour of the "white race" operates intensely against people of Negro ancestry, appreciably against those of Chinese or Japanese ancestry, and not at all against those of Indian ancestry. Thus Mr. Curtis, who was Vice-President under Mr. Hoover, had no shame in admitting Red Indian "blood." But a man with the same amount of Chinese "blood" would have had no chance of becoming Vice-President, although both on psychological tests and on cultural achievements the Chinese are vastly superior to the Red Indians. We see then that the "colour bar" does not operate very logically.

If it were established that the cultural achievements of a race were determined by its innate capacity, the Chinese would rank very high and Red Indians lower than West African Negroes as possible ancestors for a future civilized community. Further, the genes needed for cultural achievement, if they exist, are almost certainly different from those for hair shape and skin colour. In fact the existing colour bars in various countries, which are not, as a matter of fact, based on rational grounds, do not fulfil the functions which are attributed to them by their defenders. They serve to perpetuate a peculiar type of society, including not only exploited coloured people, but poor whites forced down by Negro competition to the Negro economic level, and unwilling to co-operate with the Negroes in a struggle for better conditions.

There is one racial characteristic that is of considerable interest because it is not altered by migration and its genetical determination is very simple. That is the characteristic frequencies with which various blood groups occur. Table 7 shows that in populations living side by side in Hungary the Germans closely resemble Germans in Germany, and the gypsies closely resemble the inhabitants of Northern India, speaking languages akin to the Romany language. The value of such data is that they enable us to follow the results of racial crossing in a quantitative manner. It is possible to predict what will happen as regards frequency of blood groups when two races are crossed. On the basis of that we can at once say that although the Jews may have a certain amount of ancestry of nomadic desert dwelling tribes, they resemble far more the average people living round the Mediterranean than the modern nomadic desert Arabs, some of whom have had remarkably little racial contamination from outside. To class the Jews with Arabs as a Semitic race on the basis of their language and traditions is extremely hazardous. A few data are also available regarding other serological characters.

We may sum up what little is known as to the nature of innate racial differences as follows. As regards colour or hair form, two races may differ absolutely, that is to say, there will be no overlap. As regards skeletal characters, serological characters such as membership of a blood group, and such psychological char-

176

TABLE 7

Frequencies of the Four Blood Groups in Certain
Populations

Race and Habitat	Percentage Frequencies			
	O	A	B	AB
Germans (Heidelberg)	40	43	12	5
Germans (Hungary)	41	43	13	3
Magyars (Hungary)	31	39	19	12
Gypsies (Hungary)	34	21	39	6
Indians (Northern India)	31	19	41	9

The Indians were a mixed group of soldiers, mostly from Northern India, fighting at
Salonika.

acters as we can assess, the differences are always statistical. The colour and hair form are determined by relatively few genes. The skeletal and psychological characters, as far as they are innate, are determined by many different genes. The psychological characters may be greatly influenced by culture. The serological character appears to be of no selective value. It may be that in tropical races there is selection in favour of dark skin colour and in warm and wet countries in favour of a large, broad nose. No such possibility appears to exist in the case of serological characters, which are therefore valuable for classificatory purposes just because they are trivial.

Now for the question of race crossing. How are we to evaluate these facts? I begin with a remark whose extreme simplicity shows the extraordinary lack of realism which is usually found in the discussion of this matter. When people say that racial mixture is a bad thing or a good thing they generally do not say whether they are referring to the first generation or to later generations. Now any breeder of domestic animals will at once realize the great difference involved. In poultry breeding we use the first cross between two pure races to a considerable extent. It is uniform and vigorous, and often better than either of the original races. We do not carry on further because we know that in the second generation we shall get a consider-

TABLE 8

The Nature of Interracial Differences

Characters	Absolute or Statistical Difference	Many or Few Genes	Influence of Nurture	Selective Value
Colour and hair form	A or S	F	+	?+
Skeletal characters	S	M	+	?+
Serological characters	S	F	–	–
Psychological characters	S	M	++	+?

ably greater variation, and generally a certain loss of the vigour found in the first generation of the cross. I can quite imagine that in a world under a eugenical dictatorship (if you can imagine anything so unpleasant), the large bulk of the population would be drawn from the first cross between two pure races, which would be carefully kept apart, while the first cross would not be allowed to breed further. It would be quite consistent with the belief in the importance of racial purity to hold that such an organization of the world would be desirable.

When we find undesirable or desirable qualities in mixed races we are at once up against the question of whether we are to attribute them to biological facts or to the interaction of two different cultures. It is said by some workers that the children of mixed marriages inevitably enjoy somewhat lower social standards than either of the parents because of the clash between the parental types of cultures. On the other hand one can point to some historians and sociologists who regard the contact of different cultures as an important condition of human progress. It is possible that the clash of cultures causes unhappiness for individuals but ultimately leads to social progress. With regard to the first generation from the cross, Mjoen claims that he finds a good deal of instability of character, mainly moral, in the first generation of the cross between Swedes and

Lapps. That instability might be put down to the clash of cultures. It is more usual to suggest that in later generations disharmonies arise. The statement has been made that in certain mixed populations in Africa abnormal shapes of the palate are commoner than in either pure white or pure Hottentot populations. There is the further suggestion that two different races may each of them have the genes necessary for some desirable characteristic such as resistance to a disease, but they may accomplish the same end by means of different genes, and that therefore in the second or later generation individuals may lose this resistance. There is an elementary point to be made here. The various races have the type of disease resistance which is more or less appropriate to their environment. Whatever he may be in other respects, as regards disease resistance the Englishman is a better man than a Negro in England and the Negro is a better man in West Africa, which is known as the white man's grave. It seems to me that where you have evidence of adaptation to environment like that it would be desirable to discourage not merely racial interbreeding, but emigration between the two countries except in so far as it may be necessary for a certain amount of cultural contact.

In general, however, we lack adequate evidence as to the abilities of mixed races. We find, for example, that the coloured people of the United States, who are mainly mixed, appear to do rather badly in intelligence tests, but that in India the equally mixed Anglo-Indians are sufficiently intelligent and reliable to play a very large part in running the admirable railway and telegraphic systems of their country. We have also to ask ourselves what are the possibilities of a permanent segregation of two races living in the same community, the policy which is now given legal sanction in South Africa and Australia. Perhaps the most remarkable experiment of that kind was made in India. We are now beginning to learn something of the population which lived in the valley of the Indus before the Aryan invasion.

They worshipped a mother goddess. As we have no literature from such a culture our reconstruction of it must inevitably be imaginative, and we may do worse than listen to a great poet.[1]

"Je régrette les temps de la grande Cybèle
Qu'on disait parcourir, gigantesquement belle,
Sur un grand char d'airain, les splendides cités:
Son double sein versait dans les immensités
Le pur ruissellement de la vie infinie.
L'Homme suçait, heureux, sa mammelle bénie,
Comme un petit enfant, jouant sur ses genoux.
Parcequ'il etait fort, l'Homme etait chaste et doux."

[1] Rimbaud, *Soleil et Chair*.

The people of Chanhu-daro carried their gentleness so far that, although numerous toys have been found in the ruins of their city, not a single weapon has been discovered in them. It may be that this type of culture had no future, that it was necessary for the ultimate progress of mankind that these people should be conquered by a more warlike group. If so, we can only regret it, but it is surely needless to describe the Aryans as a superior race.

It appears that these peaceful people were rapidly conquered by the Aryan invaders from the north, who regarded themselves as much superior and tried to crystallize the distinction in the caste system. That attempt was to a very large extent a failure. It is true that in any given district in India the Brahmins are probably racially different from the low caste and out-caste people, but it is nevertheless the case that in Southern India the Brahmins are pretty dark in their colour, which suggests considerable racial intermixture with the darker aboriginal inhabitants.

It is interesting to speculate whether such a policy will succeed in South Africa. My own view is that in spite of the laws in existence it is unlikely that racial mixture will be permanently prevented there. We must remember that the Europeans are in a minority. On the other hand, in Australia it is probable that this policy will end in the extinction of the aboriginal

183

blacks who are present in very small numbers. While we can imagine this policy of racial segregation combined with racial inequality succeeding in South Africa for a period to be measured in decades, or perhaps even in centuries, it is obvious that a society so constituted will be considerably less stable than a homogeneous one, and it is difficult to suppose that a period of instability will not ultimately arise in which these barriers will be swept away. However that may be I would urge the extraordinary importance of a scientific study of the effects of racial crossing for the future of the British Commonwealth. Until such a study has been accomplished, and it is a study that will take generations to complete, we are not, I think, justified in any dogmatism as to the effect of racial crossing. It may not be desirable to forbid it, but there can be very little reason, I think, to encourage it as between the widely different races of mankind. It must be remembered, however, that a policy of at any rate frowning to some extent on racial mixture, even if it were thought desirable as between the major human groups, is certainly not to be justified in Europe, if only because there are no groups within Europe that are racially pure in the sense of not overlapping with other groups as regards innate physical characteristics. Therefore, if you would forbid a fair-haired Swede to marry a dark Sicilian you would logically be compelled to forbid him from marrying a dark-haired Swede also.

I regret as much as anyone else the impossibility of coming to any reasoned conclusion on this question of racial intermixture. I feel that to forbid it may be not only impracticable in a mixed community such as South Africa, but unjust from the social or even undesirable from the biological point of view. Yet I also realize that it is an irreversible process, and may possibly be disadvantageous for the future of our species, which may demand a certain degree of specialization such as is found among different races. I am sure that the fact of our ignorance is a deplorable one which we ought to remedy.

I will now attempt to summarize these lectures, and to suggest extensions of the way of thought which I have advocated into other fields. We have surveyed in a very cursory manner the bearing of genetics on some social problems. Our results have, I think, been mainly negative. We can ascribe a certain value to negative eugenics, but we realize that negative eugenics is not confined to the suggestion that the unfit should be sterilized. We realize that there may be other more desirable ways of preventing the breeding of persons carrying undesirable dominant genes. We also realize that the discouragement of inbreeding may be a substantial part of the programme of a balanced negative eugenics. When we come to the question of positive eugenics we are more doubtful. We may think that the differential birth-rate will inevitably lead to a degeneration of the population, but yet we have to realize that this assumption, which seems not unreasonable on the basis of many statements which are commonly believed about heredity, is not borne out by the only case where we can perhaps point to historical facts. I refer to the case of polygamists as opposed to monogamists in Western Asia.

There may be something wrong with the premises that lead one to think of the differential birth-rate as likely to cause national degeneracy. If we believe that the differential birth-rate is an evil thing, if we think it is desirable that as many children as possible should be born into the more favourable environment that a relatively wealthy family can afford, we shall still find reason to question the adequacy of measures suggested for dealing with the differential birth-rate. We shall not find that family endowments in France or Belgium have done anything to increase the birth-rate in the classes to which they are given. We shall not find that Mussolini has been very successful with his measures, and we wait to learn what success is achieved by Hitler's laws and propaganda after the initial excitement of the National Socialist revolution has died down.

We must realize that the biology of human reproduction is a field of which we know singularly little. We are just beginning to learn something of its physiology and biochemistry, particularly of its chemical regulation by a series of hormones. We are learning that in animals fertility depends partly on genetical factors, but very largely on the environment. When we learn that it may be controlled not only by food but by illumination, and that this latter fact has only been established within the last five years, we shall feel sceptical as to the probable efficacy of governmental measures of control which are not based on a careful study of human biology.

Nevertheless, we shall realize that in so far as the low birth-rate in a given group is due either to late marriage or to contraception it can be controlled by the action of the community, provided that this action is based on sound economics and sound psychology. We shall, however, if we are honest, admit that the economic and psychological views of our rulers have a very slight claim to be regarded as scientifically sound, and we shall be sceptical of the efficacy of their schemes for human improvement.

In the course of our examination of theories which we could not fully accept, we came across a variety of facts. Many of those facts which I have presented to you must have seemed to you to have no bearing whatever on any practical problems. This is true at the moment; nevertheless I believe that it is only by the accumulation of facts which may at first sight appear irrelevant that we are likely to make any great progress. It may be suggested that the relation of eugenics to genetics, the scientific study of the laws of heredity and variation, is the same as the relation of astrology to astronomy. I think that would be an unfair comparison. While there is a certain amount in the existing eugenic programme of which we can wholeheartedly approve, yet we must realize that a good deal of it is the somewhat unscientific application of prejudices, whether racial or class prejudices. I would insist on the importance of human genetics and point out that there is at present exactly one professor of genetics in England. Workers within the field of genetics are mainly concerned with facts concerning plants and animals, but also concerning men. I think that they will be applied in the future, but how they will be applied it is impossible to predict. The history of astronomy furnishes a useful analogy. At first the movements of planets and stars were studied very largely with a view to astrological prediction, and many eminent astronomers made a comparatively good income by telling fortunes, which I am sure they did to the best of their ability. It was later discovered that the data accumulated with a view to astrology were of supreme importance to navigation, and it was by means of the quadrant and the chronometer that the great voyages of the eighteenth century were made. In the seventeenth century the facts of astronomy began to have an enormous influence on the general development of physics, particularly in the hands of Newton. In the nineteenth century they began to affect chemistry. It may well be, therefore, that investigations as to human heredity originally undertaken with a view to the improvement of the race may lead to other desirable ends. It may be that it will be much more important for us to know something about the innate abilities of a given child in order to be able to choose for it the type of education and career that is most suitable, than in order to encourage or discourage the production of children of some particular type. It is perfectly possible that we have sufficient potentially great men and women to fill our entire needs without the application of any eugenic measures to increase their numbers, and that a study of human innate endowment and a careful distinction between the effects of nature and nurture will enable us to discover them. It is equally possible that our existing system combs out the population pretty thoroughly for certain types of ability, though it certainly does not do so for all.

There are of course a number of other branches of human biology besides genetics. There are the medical sciences, which had an immense influence in the last century by abolishing the major epidemic diseases other than influenza, and which offer the possibility of the cleaning up of the wet tropical zones of the world which have been called the slums of our planet. It is worthy of remark that the social consequences of making the wet tropical zones healthy would be at least as great as have been the consequences of making the temperate zones healthy by ridding them of water-borne and other diseases. Medical science is now advancing towards the idea that the food supply should be standardized as is the water supply, that it is as much of a disgrace to a community that its children should be brought up on an inadequate diet as that they should drink a water supply contaminated with typhoid fever. It is conceivable that a standard food supply will be regarded as a necessity and that the supply of food will be taken over by the State. It is important to remember that the existing standards of individual needs in food are very unsatisfactory, in that they do not take sufficient account of individual variations as regards these needs. There are strains of rats that will develop rickets on a diet on which other strains do not. We must be careful not to base our standards on averages. We must realize that on a diet inadequate for the majority a certain number of children will grow up perfectly well, and that on a diet adequate for the majority another group of children will require a supplement. Our standards as regards food are still rough.

Just as we can now lay down a standard of nutrition, and point out that many of our fellows live on a diet which does not reach the standard, we can lay down standards of industrial hygiene. Our discussion of occupational risks is largely based on accidents, and on the more dramatic diseases, such as lead poisoning. We give a good deal of publicity to unfenced machinery or gassy mines. And rightly so; but an emphasis on accidents leads us to forget that the most dangerous occupation in England is cutlery grinding. The cutlery grinders have a death-rate of over three times the average of the male population, largely from lung diseases caused by the dust which they inhale.

Or, to be accurate, this was so in 1921. We do not know if it is still true, because Part IV of the Registrar General's decennial supplement on Occupational Mortality, based on the 1931 census, has not yet been published. Not enough people are interested in industrial hygiene to see that it is published within a reasonable time. And this is so quite regardless of party politics. I find members of the Labour Party as bored as Conservatives by statistics on the death-rates in different industries.

I should be interested to know whether, to take a single example, the death-rate among potters from bronchitis is still eight times that of the general population. If so, it would not be unreasonable if a certain proportion of the funds devoted to pottery research at Stoke-on-Trent were spent on research on potters rather than pots.

But while I am sure that our standards of industrial hygiene are shamefully low, it is important to realize that there is a side to this question which has so far been completely ignored. The majority of potters do not die of bronchitis. It is quite possible that if we really understood the causation of this disease, we should find that only a fraction of potters are of a constitution which renders them liable to it. If so we could eliminate potters' bronchitis by rejecting entrants into the pottery industry who are congenitally disposed to it. We are already making the attempt to exclude accident-prone workers from certain trades. The principle could perhaps be carried a good deal further.

There are two sides to most of these questions involving unfavourable environments. Not only could the environment be improved, but susceptible individuals could be excluded. Thus my father's work on divers' paralysis and caisson disease among compressed air workers led not only to the drawing up of tables for the safe ascent of divers, but to the elimination of unduly fat men, who are particularly susceptible. It must be added that at present it is generally practicable to improve the environment, while we are very rarely able to discover what types of men are susceptible. Nevertheless, in a society which was based on a knowledge of human biology it would be realized that large innate differences exist, and men would not be given tasks to which they were congenitally unsuited. We must no more forget heredity when we are trying to improve environment than we must forget environment when trying to improve heredity. A complete concentration on one side of the problem can only lead to short-sighted action.

In the long run the application of biology to social problems must depend on the ideals of the community, and the possibility which its structure offers of realizing those ideals. If we think defence against human enemies is more important than defence against microbic enemies we shall, as we do now, spend a great deal more money on it. If not, not; as in the case of the unfortunate people of Bilbao, who built sanatoria instead of trench systems, and were massacred in consequence.

But that is only one side of the question. It is a fundamental fact first clearly pointed out by Engels, though adumbrated by Rousseau, that a number of people may all desire something and act on this desire, but that the resultant of their actions may be something which none of them wish. This fact can be well illustrated in the field of human biology. In the Middle Ages most people admired the holy man, in the nineteenth century they admired the "self-made" wealthy man. But the net result of the economic system based on a genuine admiration for holiness in the past and business ability in the present was that holy men and women hardly bred at all, and "successful" men were less fertile than the economic failures in the slums. Thus so far as holiness and business success have a genetical basis they tend to be abolished in the societies which most admire them.

It is, or should be, the main task of politics to see that the resultant of individual desires does not run counter to those desires; that, for example, a sincere desire for peace should not lead to war, either by one-sided disarmament or by the piling up of huge forces which many citizens honestly believe to be needed for defence. This is a hard enough task in the economic field. I am, indeed, one of those who think it an impossible task within the framework of our present economic system.

It is perhaps even harder in the field of social biology. For the beginnings of scientific economics go back to the seventeenth century at least. And the beginnings of a scientific study of heredity were only made seventy years ago, while the modern period

194

opened with the rediscovery of Mendel's work in 1900.

If we hope to be successful in any political or social endeavour there are two prerequisites besides good will. We must examine the system with which we have to deal, and we must examine ourselves. We must find out what we take for granted in the field of social science, and then ask ourselves why we take it for granted, a much more difficult question. We must remember that the investigator, whether a biologist, an economist, or a sociologist, is himself a part of history, and that if he ever forgets that he is a part of history he will deceive his audience and deceive himself.

Index

INDEX

Peterkin, 110n.
Piedmontese, 149f.
Plato, 9, 136
polygamy, 127, 186
Poor Laws, Royal Commission on, 19
pure line, 29ff.

Queen Victoria, 88

race, 141ff., 176
  crossing, 172ff., 178ff.
  German, 151f.
  psychology, 138f., 154f.
recessive defects, 55, 90f.
reproductive index, 114ff.
retinitis pigmentosa, 51, 59, 100
Rimbaud, 182n.
Rousseau, 193
Russell, 96
Russians, 121, 149, 150, 162, 163

Scandinavia, 165f.
Schizophrenia, 83, 97
Schwesinger, 125
segregation, 42
sex-chromosomes, 62
sex-linked inheritance, 63ff.
Sjögren, 56ff.
Socialism, 14, 18, 165
South Africa, 172, 173, 174, 181, 182ff.
Soviet Union, 14, 116f., 121
Stalin, 14
Steggerda, 159f.
sterilization, 7f., 15ff., 83ff.
Suchsland, 129
Sutherland, 99

201

Books That Live

The Norton imprint on a
book means that in the
publisher's estimation it
is a book not for a single
season but for the years.

———————————

W · W · NORTON & CO · INC.
70 FIFTH AVENUE
NEW YORK